ENVIRONMENTAL SCIENCE, ENGINEERING AND TECHNOLOGY

THE NATIONAL ENVIRONMENTAL POLICY ACT (NEPA)

Environmental Science, Engineering and Technology

Additional books in this series can be found on Nova's website under the Series tab.

Additional E-books in this series can be found on Nova's website under the E-book tab.

ENVIRONMENTAL SCIENCE, ENGINEERING AND TECHNOLOGY

THE NATIONAL ENVIRONMENTAL POLICY ACT (NEPA)

MAKSIM DRUBICH
AND
ALISA MOSER
EDITORS

Nova Science Publishers, Inc.
New York

For permission to use material from this book please contact us:
Telephone 631-231-7269; Fax 631-231-8175
Web Site: http://www.novapublishers.com

Library of Congress Cataloging-in-Publication Data

The National Environmental Policy Act (NEPA) / editors, Maksim Drubich and Alisa Moser.
p. cm.
Includes index.
ISBN 978-1-61324-876-8 (hardcover)
1. Environmental policy--United States. 2. Environmental policy--United States--History. 3. United States. National Environmental Policy Act of 1969. I. Drubich, Maksim. II. Moser, Alisa.
GE180.N375 2011 363.700973--dc23 2011018639

Published by Nova Science Publishers, Inc. † New York

CONTENTS

PREFACE

Prior to the 1960s, little formal consideration was given to the potential impact of human activity on the environment. Beginning in the late 1950s and into the 1960s, the public became increasingly aware of and concerned about those impacts. During that time, Members of Congress debated the need for national policy on the environment and for an executive-level council or committee that could provide advice to the President on environmental policy issues. The statute that ultimately addressed these needs was the National Environmental Policy Act of 1969 (NEPA). This book provides an overview of the background and history of NEPA, as well as its implementation protocol for disaster response, recovery and mitigation projects.

Chapter 1- Beginning in the late 1950s and through the 1960s, Congress reacted to increasing public concern about the impact that human activity could have on the environment. A key legislative option to address this concern was the declaration of a national environmental policy. Advocates of this approach argued that without a specific policy, federal agencies were neither able nor inclined to consider the environmental impacts of their actions in fulfilling the agency's mission. The statute that ultimately addressed this issue was the National Environmental Policy Act of 1969 (NEPA, 42 U.S.C. §§ 4321-4347).

Chapter 2- The National Environmental Policy Act (NEPA) establishes environmental policies that apply to the federal government, but it is best known for imposing environmental review procedures on federal agency actions. NEPA requires agencies to review the potential environmental impacts of their projects, recording their review in a publicly available document. There are three types of environmental documents, based on the type of review: a categorical exclusion (CE), an environmental assessment

(EA), and an environmental impact statement (EIS). Tiering, programmatic, and supplemental documents are subsets of these original documents.

Chapter 3- In the aftermath of a major disaster, communities may need to rebuild, replace, or possibly even relocate a multitude of structures. When recovery activities take place on such a potentially large scale, compliance with any of a number of local, state, and federal laws or regulations may apply. For example, when older buildings must be repaired or demolished, provisions of the National Historic Preservation Act (NHPA) may need to be considered. If rebuilding will take place in a floodplain, provisions of Executive Order 11988 on Floodplain Management may apply.

Chapter 4- On April 20, 2010, an exploratory oil well in the Gulf of Mexico exploded, killing 11 people and causing the worst oil spill in American history. The oil well was on a tract leased by BP, having obtained a lease and the relevant permits from the federal government. Under relevant federal law, federal actions that may have adverse environmental effects are required to be reviewed for potential environmental harm under the National Environmental Policy Act (NEPA). This chapter will review those environmental procedures. While there are additional environmental obligations imposed on Outer Continental Shelf (OCS) drilling by other acts, this report will not review those requirements.

Chapter 5- This chapter focuses on the National Environmental Policy Act of 1969 (NEPA) and its application to offshore oil and gas activities. The paper is divided into three sections. The first section is a general discussion of NEPA as it applies to Federal agencies. The second section addresses the Department of the Interior's (DOI) NEPA regulations and portions of its Departmental Manual that relate to the implementation of those rules for oil and gas activities. The third section focuses on NEPA questions specific to the Deepwater Horizon oil spill.

In: The National Environmental Policy Act ISBN: 978-1-61324-876-8
Editors: M. Drubich and A. Moser

Chapter 1

THE NATIONAL ENVIRONMENTAL POLICY ACT (NEPA): BACKGROUND AND IMPLEMENTATION*

Linda Luther

SUMMARY

Beginning in the late 1950s and through the 1960s, Congress reacted to increasing public concern about the impact that human activity could have on the environment. A key legislative option to address this concern was the declaration of a national environmental policy. Advocates of this approach argued that without a specific policy, federal agencies were neither able nor inclined to consider the environmental impacts of their actions in fulfilling the agency's mission. The statute that ultimately addressed this issue was the National Environmental Policy Act of 1969 (NEPA, 42 U.S.C. §§ 4321-4347).

Signed into law by President Nixon on January 1, 1970, NEPA was the first of several major environmental laws passed in the 1970s. It declared a national policy to protect the environment and created a Council on Environmental Quality (CEQ) in the Executive Office of the President. To implement the national policy, NEPA required that a detailed statement of environmental impacts be prepared for all major

* This is an edited, reformatted and augmented version of Congressional Research Service Report RL33152, dated January 10, 2011.

federal actions significantly affecting the environment. The "detailed statement" would ultimately be referred to as an environmental impact statement (EIS).

With an initial absence of regulations specifying implementation procedures and no agency authorized to enforce the law, federal agencies reacted in different ways to NEPA's requirements. Some had difficulty complying with the law's EIS requirements. As a result, litigation that served to interpret NEPA's requirements and enforce agency compliance began almost immediately. In addition to questions of procedure (e.g., how, when, or why an EIS must be prepared), another question was how the environmental policy goals of the act should be implemented or enforced. The courts ultimately decided that NEPA is a procedural statute with twin aims requiring agencies to (1) consider the environmental impacts of their proposed actions and (2) inform the public that they (the agencies) considered environmental concerns in their decision-making process. In that capacity, NEPA has become a primary mechanism for public participation in the federal decision-making process.

As it has been implemented, most agencies use NEPA as an "umbrella" statute. As such, NEPA forms a framework to coordinate or demonstrate compliance with any study, review, or consultation required by other environmental laws. The use of NEPA in this capacity can lead to confusion. The need to comply with another environmental law, such as the Clean Water Act, may be identified within the framework of the NEPA process, but NEPA itself is not the source of the obligation. Theoretically, if the requirement to comply with NEPA were removed, compliance with each applicable law would still be required.

INTRODUCTION

Prior to the 1960s, little formal consideration was given to the potential impact of human activity on the environment. Beginning in the late 1950s and into the 1960s, the public became increasingly aware of and concerned about those impacts. During that time, Members of Congress debated the need for a national policy on the environment and for an Executive-level council or committee that could provide advice to the President on environmental policy issues. The statute that ultimately addressed these needs was the National Environmental Policy Act of 1969 (NEPA, 42 U.S.C. §§ 4321-4347).

Signed into law by President Nixon on January 1, 1970, NEPA was the first of several major environmental laws passed in the 1970s. It declared a national policy to protect the environment. To implement its policy, NEPA requires federal agencies to provide a detailed statement of environmental impacts, subsequently referred to as an environmental impact statement (EIS), for every recommendation or report on proposals for legislation and other major federal action *significantly* affecting the quality of the human environment.

The act also created the Council on Environmental Quality (CEQ) in the Executive Office of the President. Among other duties, CEQ provides oversight of NEPA's implementation. In 1978, CEQ was authorized by executive order to issue regulations applicable to all federal agencies regarding the preparation of EISs. However, CEQ was not authorized to enforce those regulations.

As it was subsequently interpreted, NEPA is a procedural statute with two primary aims. First, it obligates federal agencies to consider every significant aspect of the environmental impact of an action *before* proceeding with it. Second, it ensures that the agency responsible for the action will inform the public what the action is and that it has considered environmental concerns in its decision-making process. In this capacity, NEPA has become one of the primary mechanisms through which the public is able to participate in the federal decision-making process.

As a procedural statute, NEPA does not require agencies to elevate environmental concerns above others. Instead, NEPA requires only that the agency assess the environmental consequences of an action and its alternatives before proceeding. If the adverse environmental effects of the proposed action are adequately identified and evaluated, the agency is not constrained by NEPA from deciding that other benefits outweigh the environmental costs and moving forward with the action.

Most agencies use NEPA as an umbrella statute—that is, a framework to coordinate or demonstrate compliance with any studies, reviews, or consultations required by any other environmental laws. The use of NEPA in this capacity can lead to confusion. The need to comply with another environmental law, such as the Clean Water Act or Endangered Species Act, may be identified within the framework of the NEPA process, but NEPA itself is not the source of the obligation. Theoretically, if the requirement to comply with NEPA were removed, compliance with each applicable law would still be required.

Unlike other environment-related statutes, no individual agency has enforcement authority with regard to NEPA's environmental review requirements.[1] This absence of enforcement authority is sometimes cited as the reason that litigation has been chosen as an avenue by individuals and groups that disagree with how an agency meets NEPA's mandate or EIS requirements for a given project. (For example, a group may charge that an EIS is inadequate or that the environmental impacts of an action will in fact be significant when an agency claims they are not). Critics of NEPA charge that those who disapprove of a federal project will use NEPA as the basis for litigation to delay or halt that project. Others argue that litigation only results when agencies do not comply with NEPA's procedural requirements.

Environmental groups often refer to NEPA as the "Magna Carta" of environmental law. They view it as an essential tool to help agencies plan and manage federal actions in a responsible way by requiring policymakers and project sponsors to consider the environmental implications of their actions before decisions are made. Environmental groups also view the NEPA process as an important mechanism in providing the public with an opportunity to be involved in agency planning efforts. Critics charge that the law creates a complicated array of regulations and logistical delays that stall agency action.

This report provides information about NEPA's background and legislative history, provisions of the law, the role of the courts and CEQ in its implementation, how agencies implement NEPA's requirements, how the public is involved in the NEPA process, the means by which NEPA is used as an umbrella statute to coordinate or demonstrate compliance with other environmental requirements, and claims by some stakeholders that NEPA causes delays in some federal actions.

NEPA and Its History

In the 1950s and 1960s, Congress began to react to increasing public concern about the environment. In the congressional debates that ensued, a key legislative option considered was the declaration of a national environmental policy. Such a policy would require all federal agencies, whose actions were often seen as significant sources of pollution, to adhere to certain environmental values and goals. Advocates of a national policy argued that without a specific environmental policy, federal agencies were neither able nor inclined to consider the environmental impacts of their actions in fulfilling the agency's mission. Debate also existed regarding the creation of an Executive-level board or council that would gather information regarding the state of the environment and provide environmental policy advice to the President.

Background and Legislative History

For at least 10 years before NEPA was enacted, Congress debated issues that the act would ultimately address. The act was modeled on the Resources and Conservation Act of 1959, introduced by Senator James E. Murray in the 86th Congress. That bill would have established an environmental advisory counsel in the office of the President, declared a national environmental policy, and required the preparation of an annual environmental report.[2]

In the years following the introduction of Senator Murray's bill, similar bills were introduced and hearings were held to discuss the state of the environment and Congress's potential responses to perceived problems. In 1968, a joint House-Senate colloquium was convened by the chairmen of the Senate Committee on Interior and Insular Affairs (Senator Henry Jackson) and the House Committee on Science and Astronautics (Representative George Miller) to discuss the need for and potential means of implementing a national environmental policy. In the colloquium, some Members of Congress expressed a continuing concern over federal agency actions affecting the environment. Governor Laurence Rockefeller, a participant in the colloquium, stated before the joint committee:

> [W]e do not have a clearly stated national attitude toward the environment. In the areas of civil rights, education, full employment, and a number of others, the Congress of the United States has set forth a clearly understood national policy. This lack of overall national policy has been reflected in recent action of the courts in reversing decisions of administrative agencies on the grounds that they did not give sufficient consideration to environmental factors. Clearly, these agencies need better guidelines.[3]

Many of the concepts and ideas drawn from this colloquium would ultimately form the basis for the bills that would become NEPA. For example, in discussing new approaches to government, Senator Jackson argued that new approaches to environmental management were required, and he urged the colloquium to provide thoughts on possible "action-forcing" processes that could be put into operation.[4] The discussion of action-forcing processes to implement a national policy provided the seeds of the idea that would eventually become the requirement to prepare an environmental impact statement.

The bills that would become NEPA were introduced in the Senate and House in 1969 by Senator Jackson and Representative John Dingell.[5] In introducing the Senate bill, Senator Jackson stated that its purpose was to "lay the framework for a continuing program of research and study which will insure that present and future generations of Americans will be able to live in and enjoy an environment free of hazards to mental and physical well-being."[6] To accomplish this end, the Senate bill authorized federal agencies to conduct investigations and gather data on environmental issues. The bill also established a Council on Environmental Quality to analyze and study the information gathered and to advise and assist the President in the formulation of national policies.

The Senate Committee on Interior and Insular Affairs held a hearing on the proposed bills[7] in April 1969. During the hearing, the concept of creating some action-forcing mechanism, as a means of implementing a national environmental policy, was again discussed. One of the witnesses to provide testimony at the hearing was Dr. Lynton Caldwell.[8] An interchange during the hearing between Dr. Caldwell and Senator Jackson is considered by some as the point at which the provision behind the environmental impact statement requirement was introduced. Following are relevant excerpts from that testimony:

> Dr. Caldwell: I would urge that in shaping [an environmental] policy, it have an action-forcing, operational aspect ... For example, it seems that a statement of policy by the Congress should at least consider measures to require Federal agencies, in submitting proposals, to contain within the proposals an evaluation of the effect of those proposals upon the state of the environment...
>
> Senator Jackson: I have been concerned with the inadequacy of the policy declaration in the bill I have introduced. Obviously this is not enough. It does, however, provide a predicate from which to launch at a discussion as to what is required and how we should proceed ... [W]hat is needed in restructuring the governmental side of this problem is to legislatively create those situations that will bring about an action-forcing procedure the departments must comply with. Otherwise, these lofty declarations are nothing more than that.[9]

Senator Jackson further discussed the potential of broadening the policy provision in the bill to stipulate a general requirement applicable to all agencies that have responsibilities that affect the environment. In doing so, the Senator stated that he was "trying to avoid recodification of all the statutes."[10]

After the Senate hearing, Senator Jackson introduced amendments to the Senate bill.[11] Included in the amendments was a declaration of national environmental policy. Another amendment included a requirement that "all agencies of the Federal Government ... include in every recommendation or report on proposals for legislation and other major Federal actions significantly affecting the quality of the human environment, a *finding* by the responsible official that ... the environmental impact of the proposed action has been studied and considered."[12] [Emphasis added.]

In July 1969, the Senate passed its version of NEPA (S. 1075) without debate and no amendments offered. In September 1969, the House passed its version (H.R. 12549) by 372 to 15. The following December, the conference committee subsequently reported out a version containing various additions and compromises. In particular, in conference, the requirement for all major federal actions to be preceded by a "finding" on environmental impacts was changed to the requirement that a "detailed statement" on environmental impacts be prepared. The detailed statement would later be referred to as an environmental impact statement (EIS). Also included in conference was the requirement that certain federal agencies, other than the one preparing the EIS, be required to review the detailed statement.

It is unclear from the legislative history whether Congress intended for the EIS requirement to become the central element of NEPA compliance that it has. However, in addition to discussions regarding the need for action-forcing provisions to enforce the environmental policy, the legislative history includes statements regarding the need for federal agencies to consider the impacts of their actions. In discussing the relationship of the proposed legislation to existing policies and institutions, the Senate report states: "Many older operating agencies of the Federal Government ... do not at present have a mandate within the body of their enabling laws to allow them to give adequate attention to environmental values ... [The Senate bill] would provide all agencies and all Federal officials with a legislative mandate and a responsibility to consider the consequences of their actions on the environment."[13]

In late December, after minimal debate, both the House and Senate agreed to the conference report. On January 1, 1970, President Nixon signed NEPA into law.

In the more than 30 years since passage of NEPA, Congress has amended the law only to include minor technical changes.[14] However, within a year after NEPA's passage, a section was added to the Clean Air Act (42 U.S.C. 7401 et seq.) that affected the way NEPA is implemented. To further clarify agencies' responsibilities with regard to public involvement in the NEPA process, in December 1970, Congress added § 309 to the Clean Air Act.[15] Provisions of § 309 made explicit that the Administrator of the newly formed

Environmental Protection Agency (EPA) has a duty to examine and comment on all EISs. After that review, the Administrator was directed to make those comments public and, if the proposal was environmentally "unsatisfactory," to publish this finding and refer the matter to the CEQ. EPA subsequently developed a program for reviewing and rating federal agency projects (see "EPA's Unique Role in the NEPA Process" section, below).

Overview of NEPA's Provisions

The goals of NEPA are to declare a national environmental policy, provide federal agencies with action-forcing provisions intended to ensure that the goals of the policy are implemented, establish the Council on Environmental Quality (CEQ) to provide advice to the President on environmental matters and to monitor the state of the environment, and require the President to submit to Congress an annual report on the state of the environment. These provisions are contained within NEPA's two titles. Title I declares a national environmental policy that states, in part:

> [I]t is the continuing policy of the Federal Government, in cooperation with State and local governments, and other concerned public and private organizations, to use all practicable means and measures ... to create and maintain conditions under which man and nature can exist in productive harmony ... [16]

The act also specifies broad national goals. NEPA declares that it is the "continuing responsibility of the Federal Government to use all practicable means, consistent with other essential considerations of national policy, to improve and coordinate Federal plans ... [so] that the Nation may—

- Fulfill the responsibilities of each generation as trustee of the environment for succeeding generations;
- Assure for all Americans safe, healthful, productive, and aesthetically and culturally pleasing surroundings;
- Attain the widest range of beneficial uses of the environment without degradation, risk to health or safety, or other undesirable and unintended consequences;

- Preserve important historic, cultural, and natural aspects of our national heritage, and maintain, wherever possible, an environment which supports diversity, and variety of individual choice;
- Achieve a balance between population and resource use which will permit high standards of living and a wide sharing of life's amenities; and
- Enhance the quality of renewable resources and approach the maximum attainable recycling of depletable resources."[17]

Title I also includes the action-forcing, procedural requirements intended to ensure that federal agencies adhere to NEPA's goals. Section 102 forms the basic framework for federal decision making under the "NEPA process." This section includes several provisions and requires that policies, regulations, and public laws of the United States be interpreted and administered according to NEPA's policies. Among other things, Section 102 requires federal agencies to use a "systematic, interdisciplinary approach" in planning and decision making that may have an impact on the environment.

To ensure that environmental impacts are considered, Section 102(2)(C) of NEPA requires all federal agencies to include in "every recommendation or report on proposals for legislation and other major Federal actions significantly affecting the quality of the human environment, a detailed statement by the responsible official on the environmental impact of the proposed action."[18] In addition to environmental impacts, federal agencies are required to provide an analysis of

- any adverse environmental effects that cannot be avoided should the proposal be implemented,
- alternatives to the proposed action,
- the relationship between local short-term uses of man's environment and the maintenance and enhancement of long-term productivity, and
- any irreversible and irretrievable commitments of resources that would be involved in the proposed action should it be implemented.[19]

NEPA also requires federal agencies to study and develop appropriate alternatives to the recommended action for any project involving "unresolved conflicts concerning alternative uses of available resources."[20] This requirement is not limited to actions that require an environmental impact statement.

Section 103 of NEPA directs all federal agencies to review their existing statutory authority, administrative regulations, and policies and procedures to determine whether any deficiencies or inconsistencies would "prohibit full compliance with the purposes and provisions" of the act.[21] After conducting this review, the agencies are directed to take necessary measures to make their policies conform with NEPA's intent. NEPA also states that its policies and goals are supplementary to existing law.[22]

Title II of NEPA establishes the CEQ in the Executive Office of the President and specifies its responsibilities. It requires CEQ to submit to Congress an annual Environmental Quality Report on such topics as the condition of the environment, trends in the quality of the environment, and a review of federal, state and local programs to address environmental concerns.[23] Also, the act specifies a list of duties and functions that allow CEQ to support the President in information gathering and policy making with regard to environmental issues.

THE EVOLUTION OF NEPA'S IMPLEMENTATION

NEPA is a declaration of policy with action-forcing provisions, not a regulatory statute comparable to other environmental laws intended to protect air, water, wetlands, or endangered species.[24] It establishes the basic framework for integrating environmental considerations into federal decision making. However, the law itself does not provide details on how this process should be accomplished.

With an initial absence of regulations specifying implementation procedures and no agency authorized to enforce its requirements, federal agencies have reacted in different ways to NEPA's requirements. Some initially had difficulty complying with NEPA. Others believed that they were not required to comply with NEPA's provisions at all. As a result, litigation to enforce agency compliance with NEPA's mandate began almost immediately.

In addition to questions of procedure (e.g., how, when, or why an EIS must be prepared), another question ultimately to be determined was how the environmental policy goals of the act should be implemented or enforced. The courts and CEQ played significant roles in determining how those questions were answered and, consequently, how NEPA was ultimately implemented.

The Role of the Courts in Implementing NEPA

Almost since NEPA's enactment, the courts have played a prominent role in interpreting and, in effect, enforcing NEPA's requirements.[25] Beginning almost immediately and continuing into the early 1980s, the courts emphasized agency compliance with NEPA's procedural EIS requirements but did little to delineate specific compliance requirements connected to the substantive environmental policy goals. In 1983, the U.S. Supreme Court clarified that

> NEPA has twin aims. First, it places upon an agency the obligation to consider every significant aspect of the environmental impact of a proposed action. Second, it ensures that the agency will inform the public that it has indeed considered environmental concerns in its decisionmaking process. Congress in enacting NEPA, however, did not require agencies to elevate environmental concerns over other appropriate considerations. Rather, it required only that the agency take a "hard look" at the environmental consequences before taking a major action ... Congress did not enact NEPA, of course, so that an agency would contemplate the environmental impact of an action as an abstract exercise. Rather, Congress intended that the "hard look" be incorporated as part of the agency's process of deciding whether to pursue a particular federal action.[26]

This specification of NEPA's "twin aims" and the "hard look" requirement are often cited by both federal agencies and environmental advocates to articulate NEPA's mandate. In 1989, the U.S. Supreme Court reiterated that NEPA does not mandate particular results, but simply prescribes a process.[27] If the adverse environmental effects of a proposed action are adequately identified and evaluated, NEPA does not constrain an agency from deciding that other values outweigh the environmental costs. The Court further clarified that "other statutes may impose substantive environmental obligations on federal agencies, but NEPA merely prohibits uninformed, rather than unwise, agency action."[28]

In addition to determining the substantive versus procedural question, the courts have determined many specific procedural elements of NEPA compliance. For example, for individual actions, courts have ruled on agency interpretation of the meaning of the phrases "federal action," "significantly affecting," and "human environment." Also, the courts played a significant role in determining how and when federal agencies were required to prepare EISs. Some questions decided by the courts involved such issues as the adequacy of individual EISs, who must prepare an EIS, at what point an EIS must be prepared, and how adverse comments from agencies should be handled. Such decisions were, at least in part, the basis of CEQ guidelines released during the 1970s and were subsequently considered when CEQ promulgated its regulations (see "The Role of CEQ in Implementing NEPA" section, below).

The role of the courts in implementing the NEPA process has sometimes been controversial. Critics of NEPA charge that opponents of a given federal project will use litigation related to the NEPA process to delay or halt a project. Others assert that litigation is used primarily when an agency does not comply with its own NEPA procedures (see "NEPA Implementation and Project Delays" section, below).

The Role of CEQ in Implementing NEPA

Authority to promulgate regulations to implement NEPA's provisions was not expressly included among the duties and responsibilities given to CEQ under NEPA. However, shortly after signing NEPA, President Nixon issued an Executive Order authorizing CEQ to issue "regulations" for the implementation of the procedural provisions of the act.[29] The Executive Order directed CEQ to develop regulations that would be

> [D]esigned to make the environmental impact statement process more useful to decision makers and the public; and to reduce paperwork and the accumulation of extraneous background data, in order to emphasize the need to focus on real environmental issues and alternatives ... [and] require impact statements to be concise, clear, and to the point, and supported by evidence that agencies have made the necessary environmental analyses.

The Executive Order also directed federal agencies to hold public meetings and promote public involvement in the NEPA process. The Executive Order did not extend to CEQ the authority to make these regulations legally binding on federal agencies. Therefore, they would serve as only guidance for compliance.

During the 1970s, CEQ continued to issue guidelines that addressed the basic requirements of EIS preparation. CEQ left NEPA implementation largely to the discretion of federal agencies, which were to use the CEQ guidelines to prepare their own procedures. Still, many agencies were slow to do so, with many initially arguing that NEPA did not apply to them.[30]

During the mid-1970s, frequent complaints were raised regarding the delays that the NEPA process was perceived to cause in the decision-making process. Some observers attributed these problems to a lack of uniformity in NEPA implementation and uncertainty regarding what was required of federal agencies.[31] Also, in response to increasing NEPA-related litigation, agencies often produced overly lengthy, unreadable, and unused EISs.[32] In an effort to standardize an increasingly complicated NEPA process, President Carter amended President Nixon's Executive Order, directing CEQ to issue regulations that would be legally binding on federal agencies.[33] Final regulations replacing the previous guidelines were issued in the fall of 1978 and became effective on July 30, 1979.[34] CEQ's regulations were intended to foster better decision making and reduce the paperwork and delays associated with NEPA compliance.[35] CEQ's regulations also specified that the purpose of the NEPA process was to

- inform federal agencies of what they must do to comply with the procedures and achieve the goals of NEPA;
- ensure that the environmental information made available to public officials and citizens is of high quality (i.e., includes accurate scientific analysis, expert agency comments, and public scrutiny);
- foster better decision making by helping public officials make decisions based on an understanding of the environmental consequences of their actions; and
- facilitate public involvement in the federal decision-making process.[36]

CEQ's regulations, drawn in large part from its guidelines, included several noteworthy clarifications and amplifications to requirements specified in the law. For example, the regulations

- required agencies to include a project-"scoping" process to identify important environmental issues and related review requirements before writing the EIS;
- required EISs to be prepared in multiple stages (draft and final), with supplemental EISs required under specific circumstances;
- provided criteria for determining the significance of impacts and what constituted a "major federal action";

- defined and specified the roles of "lead agencies" (those responsible for preparing the NEPA documentation) and "cooperating agencies" (agencies that participate in or contribute to the preparation of the NEPA documentation);
- allowed lead agencies to set time limits on milestones in the NEPA process and page limits on documentation;
- specified a dispute resolution process between lead agencies and EPA (required originally under § 309 of the Clean Air Act), if EPA determined the EIS to be "unsatisfactory";
- specified environmental review procedures and documents applicable to projects that had uncertain or insignificant environmental impacts;
- specified how an agency was to involve the public in the NEPA process;
- required, for actions involving an EIS, that a public record of decision be published when a final agency decision is made; and
- provided for alternative compliance procedures in the event of an emergency.

The CEQ regulations were intended to be generic in nature. Each federal agency was required to develop its own NEPA procedures that would be specific to typical classes of actions undertaken by that agency.[37] Separately, CEQ regulations directed federal agencies to review their existing policies, procedures, and regulations to ensure that they were in full compliance with the intent of NEPA.[38]

CEQ's regulations are unique in several aspects. For example, they were issued eight years after enactment of the law they implement. As a result, they reflect not only CEQ's interpretation of NEPA, but also the initial interpretation of the courts and the administrative experiences of other agencies. Also, the CEQ regulations incorporated provisions of another law—§ 309 of the Clean Air Act (i.e., procedures for referring projects to CEQ for dispute resolution when EPA has found them to be environmentally unsatisfactory). Finally, although CEQ has oversight of the implementation of its regulations, it is not authorized to *enforce* them.

In addition to promulgating regulations in 1978, CEQ has provided support and informal guidance to federal agencies implementing NEPA's requirements.[39] For example, in 1981, CEQ issued its "Forty Most Asked Questions Concerning CEQ's NEPA Regulations." Answers to those questions deal with topics such as how to determine the range of alternatives considered in an EIS, how environmental documents should be made public, and the scope of mitigation measures required to be discussed. Also, in 2010, after NEPA's 40th birthday, CEQ proposed guidance intended to "Modernize and Reinvigorate NEPA."[40] Included was draft guidance for public comment on when and how federal agencies must consider greenhouse gas emissions and climate change in their proposed actions. CEQ had been asked to provide guidance on this subject informally by federal agencies and formally by a petition under the Administrative Procedure Act. The draft guidance was intended to explain how federal agencies should analyze the environmental impacts of greenhouse gas emissions and climate change when they describe the environmental impacts of a proposed action under NEPA.

Determining when NEPA Applies

Under NEPA, an environmental impact statement must be prepared for "every recommendation or report on proposals for legislation and other major federal actions significantly affecting the quality of the human environment." Interpretation of each element of this phrase has been the subject of myriad court decisions and guidance from CEQ. Two terms of particular relevance are "federal action" and "significantly." In determining whether and how NEPA will apply to an action, it is necessary to determine whether an action is in fact a federal one and, if so, whether its environmental impacts will be significant.

Federal Actions Subject to NEPA

To determine whether NEPA applies to an action, it is first necessary to determine whether it is a federal one.[41] "Federal" actions include those that are potentially subject to federal control and responsibility. Such actions include "projects and programs entirely or partly funded, assisted, conducted, regulated, or approved by federal agencies."[42] Specifically, federal agency compliance with NEPA may be required for actions that require a federal permit or other regulatory decision to proceed.[43]

In many cases, it is immediately apparent that a project or program is federal (as opposed to a strictly private or state action[44]). However, in some instances, such as private actions in which a federal agency has some small involvement, a determination is not as clear.[45] CEQ regulations specify categories of actions within which NEPA-covered federal actions tend to fall (see Table 1).

Table 1. Typical Categories of "Federal Actions" Subject to NEPA

Category of action	Examples
Site-specific projects	Construction or management activities located in a defined geographic area; actions requiring federal licensing, permitting, or other regulatory decision; activity requiring federal assistance or funding.
Adoption of programs	Projects that may include groups of concerted actions to implement a specific policy or plan; systematic and connected agency decisions allocating agency resources to implement a specific statutory program or executive directive.
Adoption of plans	May include official documents prepared or approved by federal agencies which guide or prescribe alternative uses of federal resources, upon which future agency actions will be based (agencies may argue that certain plans do not fit the definition of an "action" in accordance with NEPA).
Adoption of policy	May include the adoption of official policy, such as rules, regulations, and interpretations adopted pursuant to the Administrative Procedure Act (5 U.S.C. § 551 et seq.); treaties and international conventions or agreements (also, Executive Order 12114, Environmental Effects Abroad of Major Federal Actions, specifies environmental review requirements for actions taken outside the United States); or formal documents establishing an agency's policies that will result in or substantially alter agency programs.

Source: 40 C.F.R. § 1508.18(b).

A broad NEPA review may be done for the adoption of programs, plans, or policies. Such a review would most likely be followed by a site-specific review for any subsequently implemented projects. This process of producing a broad statement, followed by a more narrowly focused NEPA analysis is referred to as "tiering."[46] In such a project, the NEPA documentation need only summarize the issues discussed in the broader document and incorporate previous discussions by reference. Such a process is recommended to avoid repetitive discussion of the same issues. See http://www.serconline.org/SEQA/stateactivity.html.

Table 2 lists examples of projects at selected agencies that may fall into one of the categories of actions that require environmental review under NEPA.

Table 2. Selected Examples of Agency Actions Requiring Environmental Review Under NEPA

Agency	Project types
Department of Agriculture	Forest Service—Private timber, grazing, or mining operations on Forest Service land.
Department of Defense	Army Corps of Engineers—Flood control projects, ecosystem restoration projects, water resources projects, or projects requiring a federal permit for dredge and fill operations.
Department of Energy	Approval of dam construction, the process of siting of oil and gas pipelines on federal land, the process of siting power transmission lines on federal land, and research operations.
Department of Housing and Urban Development	Construction of affordable housing projects; certain projects that would remove, demolish, convert, or substantially rehabilitate existing housing units; or the extension of urban development grants or block grant programs.
Department of the Interior	Bureau of Land Management—Private mining operations on federal land; and oil and gas drilling operations on federal lands.
Department of Transportation	Federal Highway Administration—Highway and bridge construction, maintenance, and repair.
	Federal Aviation Administration—Airport construction and expansion.
EPA	Issuance of permits under the National Pollutant Discharge Elimination System (NPDES).

Source: Congressional Research Service (CRS) review of individual agency actions.

Although such application is rare, NEPA also is intended to apply to agency proposals for federal legislation. CEQ's definition of legislation includes "a bill or legislative proposal to Congress developed by or with the significant cooperation and support of a Federal agency.... "[47] This definition does not include requests for appropriations. The test for "significant cooperation" is whether the proposal is predominantly that of the agency rather than another source. Only the agency with primary responsibility for the subject matter involved is required to prepare a legislative EIS.

Determining the Significance of a Federal Action

The requirement to prepare an EIS depends on whether the federal action will have impacts "significantly affecting the quality of the human environment." In the years after NEPA was enacted, a question that was often disputed between federal agencies and third parties, and ultimately decided by the courts, was whether a given federal action had a "significant" impact. Most federal actions have *some* impact on the environment. Determining the degree of impact is necessary to determine how to comply with NEPA's procedural requirements.

CEQ regulations do not list specific types of projects that have significant environmental impacts or definitively define "significantly." Instead, the regulations require agencies to determine the significance of a project's impacts on a case-by-case basis, based on its context and intensity.[48]

Determining the context of a project involves analyzing the significance of its impacts to society as a whole, an affected region, affected interests, or the locality.[49] A site-specific project may require analysis of the local significance of the project, whereas a programmatic action may have nationwide significance. The degree of significance may depend on factors such as the location and scope of the project. For example, the impacts of a site-specific project on 1 acre of a 2,000- acre wetland may be insignificant compared with a project that affects 1 acre of a 2-acre wetland.

Intensity refers to the severity of a project's impacts. Factors used to assess an impact's intensity must be determined on a case-by-case basis. However, CEQ specifies the following minimum factors that must be evaluated:

- Environmental impacts that may be beneficial *and* adverse.
- The degree to which the proposed action affects public health or safety.
- Unique characteristics of the geographical area, such as proximity to historic or cultural resources, park lands, prime farmlands, wetlands, wild and scenic rivers, or ecologically critical areas.
- The degree to which the effects on the quality of the human environment are likely to be highly controversial (in this context, "controversy" relates only to the interpretation of the environmental effects of a project—not to the potential controversy or unpopularity of the project as a whole).

- The degree to which the possible effects on the human environment are highly uncertain or involve unique or unknown risks.
- The degree to which the action may establish a precedent for future actions with significant effects.
- Whether the action is related to other actions with individually insignificant but cumulatively significant impacts.
- The degree to which the action may adversely affect resources listed in, or eligible for, the National Register of Historic Places or may cause loss or destruction of significant scientific, cultural, or historical resources.
- The degree to which the action may adversely affect an endangered or threatened species or its habitat.
- Whether the action threatens a violation of federal, state, or local law or requirements imposed for the protection of the environment.[50]

Individual agencies may consider additional factors based on environmental impacts common to the types of projects pursued by that agency. Further, to adequately determine an impact's intensity, input from other agencies may be needed. For example, if a highway will cut across prime farmland, the Department of Transportation (DOT) may need assistance from the U.S. Department of Agriculture (USDA) to determine the intensity of the project's impacts.

Because degrees of impact must be evaluated to determine project significance, such an evaluation may require subjective judgments. Therefore, a clear administrative record is generally considered necessary to demonstrate that an agency appropriately determined the significance of a project's impacts.

Overview of the NEPA Process

The NEPA process includes the steps a federal agency must take to document that consideration was given to determine the significance of a proposed action's environmental impacts. For some actions, it may not be readily apparent that environmental impacts will be significant. Some projects clearly have little or no significant impact, but they still require an agency to demonstrate that the level of impacts was considered. To account for this variability, CEQ regulations establish the following three classes of action, which determine how compliance with NEPA analysis is documented:

- Actions Requiring an EIS—When it is *known* that the action will have a significant environmental impact.
- Actions Requiring an Environmental Assessment (EA)—When the significance of environmental impacts is *uncertain* and must be determined.
- Actions that are Categorically Excluded—Those which normally do not individually or cumulatively have a significant effect on the human environment.

The requirement to produce an EIS is probably the most familiar element of NEPA compliance. However, actions requiring an EIS account for a small percentage of all federal actions proposed in a given year. For example, in 2008, 543 EISs were filed with EPA (this total includes draft, final, and supplemental EISs).[51] CEQ estimates that the vast majority of federal actions require an EA or are categorically excluded from the requirement to prepare an EA or EIS.

Determining the total number of federal actions subject to NEPA is difficult, as most agencies track only the number of actions requiring an EIS. Also, as indicated in the figures above, agencies track the total draft, final, and supplemental EISs filed in a given year, not the total number of individual federal actions requiring an EIS in a given year. One agency that does track all projects is DOT's Federal Highway Administration (FHWA). According to FHWA, in 2007, approximately 4% of all highway projects required an EIS, 4% required an EA, and almost 92% were classified as categorically excluded. Although projects requiring an EIS or EA represent a relatively low percentage of all projects, those projects represented a comparatively large percentage of total FHWA funding. For example, while projects requiring an EIS or EA represented 8% of the total number of projects, they accounted for almost 23% of the funds allocated by FHWA in 2007. Projects classified as categorical exclusions accounted for almost 77% of FHWA funds.[52]

Environmental Impact Statements

As soon as practicable after its decision to prepare an EIS, the agency preparing it (the "lead agency") is required to publish a notice of intent (NOI) in the *Federal Register*.[53] The NOI acts as the formal announcement of the project to the public and to interested federal, state, tribal, and local agencies.

As soon as possible after, or in conjunction with, the determination that an EIS is needed, the agency is required to determine the scope of the project. During the scoping process the "lead agency" must

- identify and invite the participation of affected parties, including federal, state, or local agencies or Indian tribes; proponents of the actions; and other interested persons;
- identify significant issues to be analyzed in depth in the EIS;
- identify and eliminate issues that are *not* significant or have been covered by prior environmental review from detailed study;
- allocate assignments for preparing the EIS to relevant agencies; and
- identify other environmental review and consultation requirements so that analyses and studies required other under federal, state, local or tribal laws may be prepared concurrently, rather than sequentially, with the EIS.[54]

During the scoping process, the lead agency may set time and page limits for an individual EIS. The agency should also determine any environmental laws, regulations, or executive orders, in addition to NEPA, that may apply to the project. For example, the agency should determine in the scoping process whether the project may affect property of historical significance, endangered species habitat, or wetlands or navigable waters—each of which may require compliance with the National Historic Preservation Act, the Endangered Species Act, or the Clean Water Act, respectively.

Once the scope of the action has been determined, EIS preparation can begin. Preparation is done in two stages, resulting in a draft and a final EIS.[55] The draft EIS should be prepared in accordance with the scope of the project and, to the fullest extent possible, meet requirements of § 102(2)(C) of NEPA. The final EIS should respond to any participating agency comments and address any inadequacies in the draft EIS. A supplemental EIS may be required in some instances. A summary of the components of an EIS, as required under CEQ's regulations, is provided in Table 3.

Table 3. Components of an EIS

EIS Component	Description
Purpose and Need Statement	A brief statement, developed by the lead agency, specifying the underlying purpose of a project and the need to which the agency is responding (40 C.F.R. § 1502.13).
Alternatives	A discussion of the range of alternatives, including the proposed action, that will meet the project's purpose and need. The discussion should explore and objectively evaluate all "reasonable" alternatives, and for alternatives which were eliminated from detailed study, a brief discussion of the reasons for their having been eliminated. A "no action" alternative may also be required to establish a baseline against which other alternatives may be compared (40 C.F.R. § 1502.14).
Affected Environment	A succinct description of the environment of the area(s) to be affected by the alternatives under consideration. For example, the affected environment may include wetlands, prime farmland, urban areas, historic sites, or endangered species habitat (40 C.F.R. § 1502.15).
Environmental Consequences	An analysis of impacts of each alternative on the affected environment, including a discussion of the probable beneficial and adverse social, economic, and environmental effects of each alternative. This section must also include, where applicable, a discussion of both the direct and indirect effects of each alternative and the significance of those effects; a description of the measures proposed to mitigate adverse impacts; and methods of compliance with any applicable legal requirements may (e.g., whether and how compliance with the Endangered Species Act will be accomplished if endangered species habitat is impacted) (40 C.F.R. § 1502.16).
List of Preparers	List of names and qualifications of individuals responsible for preparing the EIS (40 C.F.R. § 1502.17).
Appendix	Any material prepared in connection with the EIS. Such materials normally consist of material which substantiates any analysis fundamental to the EIS (40 C.F.R. § 1502.18).

Source: CRS review of requirements in 40 C.F.R. § 1502.

The action's purpose and need statement is the foundation on which subsequent sections of the EIS are built. No hard-and-fast regulatory definition of "purpose and need" exists. However, as it has been interpreted, the statement cannot be so narrow that it effectively defines competing "reasonable alternatives" out of consideration. The "purpose" of an action may be a discussion of the goals and objective of an action. The "need" may be a discussion of existing conditions that call for some improvement.

The goals defined in the purpose and need evaluation facilitate the development of viable project alternatives. CEQ regulations refer to the alternatives section of the EIS as the "heart" of the document.[56] Alternatives that must be considered include those that are practical and feasible from a technical, economic, and common-sense standpoint, rather than simply desirable from the standpoint of the agency or a potentially affected stakeholder.[57] Large, complex projects may have a large number of reasonable alternatives. In this case, CEQ suggests that only a representative number of the most reasonable examples, covering the full range of alternatives, should be presented.[58]

Once the final EIS is approved and the agency decides to take action, the lead agency must prepare a public record of decision (ROD). CEQ regulations specify that the ROD must include a statement of the final decision, all alternatives considered by the agency in reaching its decision, and whether all practicable means to avoid or minimize environmental harm from the selected alternative have been adopted and, if not, why they were not.[59]

Generally, once the ROD has been issued, an agency's action may proceed (as long as other statutory requirements are met). In addition to the EIS and the ROD, the final procedural record of the NEPA process may include, but is not limited to, planning documents, notices, scoping hearings, documents supporting findings in the EIS, public comments, and agency responses.[60]

Environmental Assessments

If an agency is uncertain whether an action's impacts on the environment will be significant, it usually prepares an environmental assessment (EA). An EA is carried out to clarify issues and determine the extent of an action's environmental effects. CEQ regulations define an EA as a concise public document that (1) provides sufficient evidence and analysis for determining whether to prepare an EIS or a finding of no significant impact (FONSI), (2) aids agency compliance with NEPA when no EIS is required, and (3) facilitates preparation of an EIS when one is necessary.[61]

The CEQ regulations require no standard format for EAs; however, the regulations do require agencies to include a brief discussion of the need for the proposal, alternatives, impacts of the proposal and alternatives, and a list of agencies and individuals consulted.[62] Individual agency regulations and guidance may include more specific requirements. Some agencies suggest that the process for developing an EA should be similar to developing an EIS. For example, the applicant should consult interested agencies to scope the project to determine the potential for social, economic, or environmental impacts; briefly discuss the project's purpose and need; identify project alternatives and measures to mitigate adverse impacts; and identify any other environmental review requirements applicable to the project (e.g., permitting requirements under Section 404 of the Clean Water Act). Public participation in the EA process is left largely to the discretion of the lead agency.

If at any time during preparation of the EA, a project's impacts are determined to be significant, EIS preparation should begin. If the impacts are determined not to be significant, the lead agency must prepare a FONSI. The FONSI serves as the agency's administrative record in support of its decision regarding a project's impact. The FONSI also must be available to the public.[63]

Categorical Exclusions

If a project is of a type or in a category known to have no significant environmental impacts, it is categorically excluded from the requirement to prepare and EA or EIS. Individual agencies are required to specifically list, in their respective NEPA regulations, those projects that are likely to be considered categorical exclusions (CEs).[64] For example, DOT has identified the construction of bicycle and pedestrian lanes, landscaping, and the installation of traffic signals as actions that would generally be classified as categorical exclusions.[65]

Whether or what types of documentation may be required to demonstrate that a project is categorically excluded will depend on whether the project involves extraordinary circumstances that may cause a normally excluded action to have a significant environmental effect.[66] An individual agency's NEPA requirements may specify criteria under which otherwise excluded actions may require documentation to prove that the CE determination is appropriate.

Although categorically excluded projects do not have significant environmental impacts, an agency may require a certain level of documentation to prove that the CE determination is appropriate. Also, the fact that a project does not have a significant impact, as defined under NEPA, does not mean that it will not trigger statutory requirements of other environmental laws. For example, if historical sites, endangered species habitats, wetlands, or properties in minority neighborhoods, to name a few, are affected by a proposed federal action, compliance with related environmental laws, in addition to NEPA, may be required.

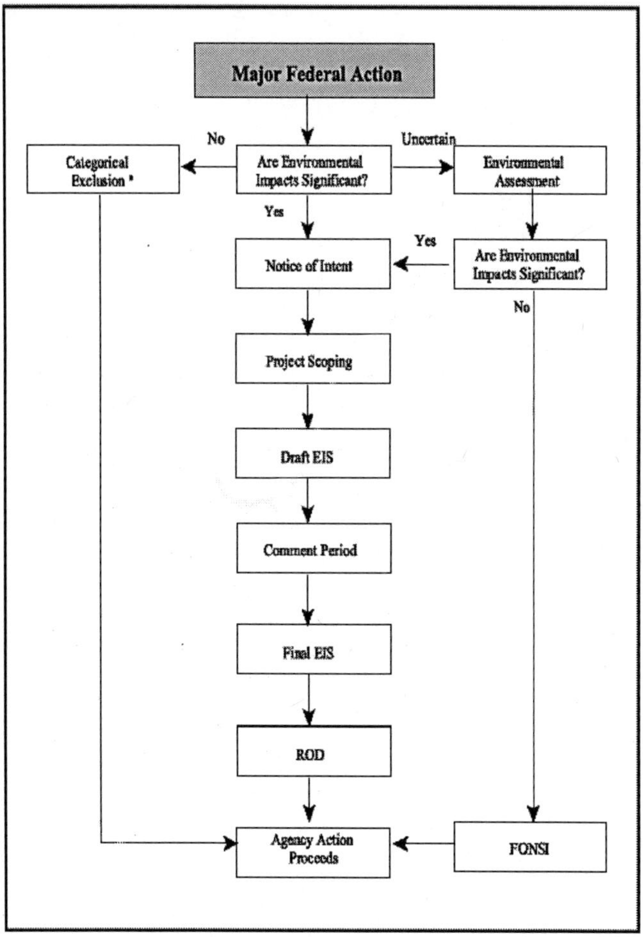

Source: 40 C.F.R. §§ 1501-1506.

[a] If an action is not specifically identified as a categorical exclusion in the respective agency's NEPA regulations, it may still require an environmental assessment to confirm that its impacts are not significant. An agency may also choose to prepare an environmental assessment, even though it is not required to do so, to aid in its compliance with NEPA.

Figure 1. Overview of the NEPA Process.

In 2010, CEQ released "Guidance Clarifying Use of Categorical Exclusions."[67] This guidance was designed to ensure that agencies establish and use categorical exclusions appropriately and transparently. It also calls on agencies to review their existing categorical exclusions periodically to avoid the use of outdated NEPA procedures.

A simplified overview of the NEPA process is illustrated in Figure 1.

AGENCY PARTICIPATION IN THE NEPA PROCESS

Federal actions to which NEPA applies involve the participation of a "lead agency" and "cooperating agencies." As stated previously, the lead agency is the federal agency that takes responsibility for preparing the NEPA documentation.[68] State or local agencies may act, with the federal lead agency, as joint lead agencies. The project applicant[69] may initially develop substantive portions of the environmental document; however, the lead agency is responsible for its scope and overall content. A cooperating agency is any federal agency, other than a lead agency, that has jurisdiction by law or special expertise regarding any environmental impact involved in a proposal.[70] A tribal, state, or local agency may also be a cooperating agency. Table 4 lists selected statutes that may apply to a given federal action and the corresponding agency that could subsequently be required to participate in the NEPA process.

Responsibilities of the Lead and Cooperating Agencies

At the request of the lead agency, the cooperating agency is required to assume responsibility for developing information and preparing environmental analyses, including portions of the EIS related to its special expertise. Such a role may be set out in a memorandum of understanding or agreement between the agencies. A cooperating agency may be excused from some or all of these responsibilities if precluded by other program requirements.

Table 4. Selected Federal Statutes and Potential Corresponding Key Cooperating Agencies

Statute	Potential cooperating agency
National Historic Preservation Act	Advisory Council on Historic Preservation and/or state or tribal historic preservation officer
Endangered Species Act	The Department of the Interior's U.S. Fish and Wildlife Service and/or the Department of Commerce's National Marine Fisheries Service
Clean Water Act	The Army Corps of Engineers and/or EPA
Wild and Scenic Rivers Act	The agency responsible for managing the listed or study river (e.g., the National Park Service, U.S. Fish and Wildlife Service, Bureau of Land Management, or Forest Service)
Statute	Potential cooperating agency
Farmland Protection Policy Act	The Natural Resources Conservation Service of USDA

Source: Table prepared by CR and based on the likely applicability of selected federal statutes.

Note: Tribal, state, and local agencies may also be included among those required to participate in a given EIS.

Some projects have involved disagreements regarding the authority of and extent to which coordinating agencies should be involved in the NEPA process. For example, some stakeholders have expressed confusion regarding the degree to which a coordinating agency has the right to influence the development of certain elements of an EIS. In 2003, this issue of agency authority was the subject of correspondence between Transportation Secretary Norman Mineta and CEQ Chairman James Connaughton. Secretary Mineta asked for clarification regarding the role of lead and cooperating agencies with regard to developing purpose and need statements.[71] Secretary Mineta referred to the sometimes extended interagency debates over purpose and need statements as a cause of delay in highway project development. In his response, Chairman Connaughton referred to CEQ regulations specifying that the lead agency has the authority and responsibility to define a project's purpose and need. Further, Chairman Connaughton referenced previous federal court decisions giving deference to the lead agency in determining a project's purpose and need. Chairman Connaughton's letter also quotes CEQ's regulations, citing the lead agency's "responsibilities throughout the NEPA process for the 'scope, objectivity, and content of the entire statement or of any other responsibility' under NEPA."

Addressing Agency Comments

Before completing an EIS, the lead agency is required to consult with and obtain comments from cooperating agencies regarding any environmental impact involved in the proposed action.[72] The CEQ regulations specify requirements for inviting and responding to comments on the draft EIS.[73] In addition to the cooperating agencies, which must comment, the lead agency is required to request comments from appropriate state, local, or tribal agencies; the public, particularly those persons or organizations who may be interested in or affected by the action (see further discussion under the "Demonstrating Public Involvement" section, below); any agency that has requested to receive EISs on similar actions; and the applicant (if there is one).[74]

If a lead agency receives comments on a NEPA document, the agency is required to assess and consider those comments and respond in one or more of the following ways:

- Modify proposed alternatives, including the proposed action.
- Develop and evaluate alternatives not previously considered.
- Supplement, improve, or modify its analyses.
- Make factual corrections in the EIS.
- Explain why the comments do not warrant further response from the lead agency, citing the sources, authorities, or reasons that support the agency's position and, if appropriate, indicate circumstances that would trigger agency reappraisal or further response.[75]

Under CEQ regulations, lead agencies are required to invite comments on a draft EIS, cooperating agencies have a duty to comment on it, and lead agencies are required to respond to those comments. As illustrated in the choices listed above, the lead agency is not precluded from moving forward with a project if it sufficiently addresses those comments. However, if negative comments are received, to avoid a potential legal challenge after the project has reached an advanced stage of development, the lead agency is well-served to resolve the issue.

EPA's Unique Role in the NEPA Process

Independent of its potential to participate as a lead or cooperating agency,[76] EPA has two distinct roles in the NEPA process. The first regards its duty, under § 309 of the Clean Air Act, to review and comment publicly on the environmental impacts of proposed federal activities, including those for which an EIS is prepared. After conducting its review, EPA must rate the adequacy of the EIS and the environmental impact of the action.[77] The EIS may be rated "adequate," "needs more information," or "inadequate." The lead agency is required to respond appropriately, depending on EPA's rating. With regard to rating the environmental impacts of an action, EPA would rate a project in one of the following four ways: lack of objections, environmental concerns, environmental objections, or environmentally unsatisfactory. If it determines that the action is environmentally unsatisfactory, EPA is required to refer the matter to CEQ for dispute resolution.[78] However, such referral should be made only after concerted, timely, but unsuccessful attempts to resolve differences with the lead agency.[79]

EPA's second duty is an administrative one, in which it carries out the operational duties associated with the EIS filing process. In 1978, these duties were transferred to EPA by CEQ in accordance with terms of a Memorandum of Agreement (MOA).[80] Under the MOA, EPA's Office of Federal Activities is designated the official recipient of all EISs prepared by federal agencies. EPA maintains a national EIS filing system. By maintaining the system, EPA facilitates public access to EISs by publishing weekly notices in the *Federal Register* of EISs available for public review, along with summaries of EPA's comments.

DEMONSTRATING PUBLIC INVOLVEMENT

As the law has been interpreted, one of the primary goals of NEPA is to give the public a meaningful opportunity to learn about and comment on the proposed actions of the federal government *before* decisions are made and actions are taken. To meet this goal, CEQ's regulations require agencies to encourage and facilitate public involvement in decisions that significantly affect the quality of the human environment (i.e., projects that require an EIS).[81] Specifically, agencies are required to provide public notice of NEPA-related hearings, public meetings, and the availability of environmental documents so as to inform public stakeholders that may be interested in or affected by a proposed action.[82] Documentation related to the public's participation in the NEPA process (e.g., public comments or hearings transcripts) must be included in the final EIS.

As mentioned above, the lead agency must seek and respond to public comments. Public stakeholders likely to comment on federal actions will vary according to the action. They may include individuals or groups expected to benefit from or be adversely affected by the project, or special interest groups with concerns about the project's environmental impacts. For example, a road-widening project may have an impact on adjacent homes or businesses. Such a project may elicit comments from the local business community (e.g., individual businesses, the Chamber of Commerce, or local development organizations) and area home owners. A project with impacts on sensitive environmental resources, such as wetlands or endangered species, may generate comments from environmental interest groups.

If stakeholders have concerns about a project's impacts, their comments may be directed at virtually any element of that project, the NEPA process, or related documentation. If stakeholder comments are not addressed sufficiently, stakeholders may respond by filing suit. To avoid conflict after a project has reached an advanced stage of development, CEQ recommends that continuous contact with nonagency stakeholders be maintained throughout the decision-making process—from the earliest project planning stages to the selection of a particular alternative, including the intervening stages to define purpose and need and to develop a range of potential alternatives. The need for such contact was illustrated in a 1997 CEQ study. Study results found that one element of the NEPA process critical to effective and efficient implementation was "... the extent to which an agency takes into account the views of the surrounding community and other interested members of the public during its planning and decisionmaking process."[83]

CEQ regulations specify public involvement requirements only for federal actions requiring an EIS. Agencies may devise their own policy regarding public involvement in the preparation of an EA or in making a categorical exclusion determination.

THE USE OF NEPA AS AN "UMBRELLA" STATUTE

Large, complex actions, such as bridge and highway construction, mining operations, or oil and gas development on public lands, may require compliance with literally dozens of federal, state, tribal, and local laws.

Depending on the resources present at a project site, compliance with various categories of legal requirements may apply to a given federal action, as illustrated in Table 5.

Table 5. Categories of Legal Requirements Potentially Applicable to Federal Actions

Category of laws	Selected potential corresponding authorities
Laws intended to protect physical resources	Clean Air Act
	Clean Water Act
	Pollution Prevention Act
	Safe Drinking Water Act
Laws intended to protect cultural resources	National Historic Preservation Act
	Archaeological and Historical Preservation Act
	Historic Sites and Buildings Act
Laws intended to protect natural resources	Endangered Species Act
	Marine Mammal Protection Act
	Fish and Wildlife Coordination Act
	Migratory Bird Treaty Act
Laws intended to minimize impacts to communities or individuals	Emergency Planning and Community Right to Know Act
	Farmland Protection Policy Act
	Title VI of the Civil Rights Act
Special status land use laws	Wild and Scenic Rivers Act
	The Coastal Zone Management Act
	Wilderness Act

Source: Table prepared by CRS and based on a review of selected federal actions.
Note: Tribal, state, or local laws may also be applicable to a given impacted resource.

To integrate the compliance process and avoid duplication of effort, NEPA regulations specify that, to the fullest extent possible, agencies must prepare the EIS concurrently with *any* environmental requirements.[84] The EIS must list any federal permits, licenses, and other entitlements required to implement the proposed project. In this capacity, NEPA functions as an "umbrella" statute; any study, review, or consultation required by any other law that is related to the environment should be conducted within the framework of the NEPA process.

NEPA forms the framework to coordinate and demonstrate compliance with these requirements. NEPA itself does not *require* compliance with them. Theoretically, if the requirement to comply with NEPA were removed, compliance with each applicable law would still be required. The use of NEPA as an umbrella statute can lead to confusion in this regard. For example, consider a project alternative that requires compliance with the Endangered Species Act (ESA). One required element of the EIS may include a demonstration that, among other potential requirements, a biological assessment be prepared in compliance with the ESA. The requirement to comply with the ESA, including the involvement of the appropriate agency with jurisdiction over compliance, would simply be *identified* within the framework of the NEPA process, not required by NEPA.

NEPA Implementation and Project Delays

Stakeholders such as state and local project sponsors and industry representatives with an interest in the implementation of a federal action sometimes charge that NEPA implementation is inefficient and overly time-consuming, leading to what they perceive as unnecessary delays in needed government actions. Some agency representatives feel that the NEPA process, when implemented as required by the CEQ regulations, actually facilitates a more efficiently executed project. Environmental organizations look at the NEPA process as a necessary step in ensuring that the public gets a voice in the federal decision-making process and that expediting that process is not necessarily in the best interest of the public or the environment. Further, they argue that blaming the environmental compliance process for project delays is misplaced. One argument is that federal projects may be delayed because resource agencies, required by law to participate in the compliance process, are overburdened and not sufficiently funded, staffed, or equipped to meet the demand.

Causes of Project Delays Attributed to the NEPA Process

Delays attributed to the NEPA process fall into two broad categories—those related to the time it takes to complete required documentation and delays resulting from NEPA-related litigation.

In the past, particularly in the years after NEPA was implemented, the preparation of NEPA documentation played a role in delaying individual federal actions. However, there is little data available to demonstrate that NEPA currently plays a significant role in delaying federal actions. This lack of data is attributable to the fact that, other than the Department of Energy and DOT, federal agencies do not routinely maintain information on the time it takes to complete the NEPA process. Therefore, gathering accurate data on how long it takes to prepare NEPA documentation, and whether the NEPA process is directly the cause of project delays, is difficult. For example, the preparation of NEPA documentation is generally done concurrently with preliminary project design. If a project undergoes specification changes, those alterations may necessitate modifications to the NEPA documentation. Consequently, the time to complete the NEPA process may be extended.

The perception that NEPA results in extensive delays and additional costs to the successful delivery of certain federal projects can be magnified when compliance with multiple environmental laws and regulations is required (see "The Use of NEPA as an "Umbrella" Statute," above). The sometimes extensive reviews, documentation, and analysis required by agencies such as the Army Corps of Engineers, the U.S. Fish and Wildlife Service, the Coast Guard, and EPA, as well as various state regulatory and review agencies, add further to the perception that extensive delays are related to the NEPA process. Such "delays" may actually stem from an agency's need to complete a permit process or analyses required under separate statutory authority (e.g., the Clean Water Act or Endangered Species Act), over which the lead agency has no authority.

Litigation is probably the most often cited cause of NEPA-related project delays. Although this may have been the case in the past, the total number of NEPA-related cases in the past 10 years has been small (especially when compared with the total number of federal actions requiring some environmental review under NEPA). For example, in 2005, a total of 118 NEPA-related cases were filed. Of those, 43 resulted in an injunction. The majority of cases were filed against two agencies—the USDA's Forest Service (with 50 cases files) and the Department of the Interior's Bureau of Land Management (with 12 cases files).[85] The main reason that plaintiffs filed suit was because they believed that the EIS or EA was inadequate (e.g., information was incomplete or the document did not sufficiently analyze the cumulative or indirect effects of an action).

NEPA litigation began to decline in the mid 1970s and has remained relatively constant since the late 1980s.[86] This trend may be due in part to improved agency compliance with promulgated regulations and improved agency expertise in preparing required documentation. However, another factor may be the decrease in the number of federal actions funded by Congress that would be defined as "major federal actions" under NEPA.[87]

Although litigation has decreased, agency concern regarding the threat of litigation may still affect the NEPA process, particularly for complex or controversial projects. In addition to CEQ regulations and an agency's own regulations, a project sponsor may be mindful of previous judicial interpretation when preparing NEPA documentation in an attempt to prepare a "litigation-proof" EIS. CEQ has observed that such an effort may lead to an increase in the cost and time needed to complete NEPA documentation, but not necessarily an improvement in the quality of the documents ultimately produced.[88]

Studies Into NEPA's Effectiveness and Causes of Delays

In the past 20 years, numerous surveys and reports, conducted by both public agencies and private organizations, have studied the effectiveness of the NEPA process. They sought to determine issues such as how the NEPA process is implemented at individual agencies, whether the NEPA process delays project implementation, and, if so, how those delays may be addressed and NEPA more effectively implemented.

In 2004, a survey of staff from the Department of Defense, the Department of the Interior, and the Forest Service sought to determine the degree to which the NEPA process slowed decision making and delayed projects.[89] The survey identified the following primary reasons for project delays:

- Decision maker changes in the project.
- Court challenges to a project.
- Poor documentation that needed to be redone.
- Changes in or additions to project alternatives.
- Compliance requirements of the Endangered Species Act.

Depending on the agency responding, factors "outside the NEPA process" were identified as the cause of delay between 68% and 84% of the time.

In 1997, CEQ published a study to determine NEPA's effectiveness and methods to improve its implementation.[90] Study participants included individuals and organizations that were knowledgeable about NEPA and could be characterized as both supporters and critics of NEPA. Generally, participants felt that NEPA's enduring legacy was that it provided a framework for collaboration between federal agencies and those who will bear the environmental, social, and economic impacts of agency decisions. However, they also felt that NEPA often takes too long and costs too much, agencies make decisions before hearing from the public, documents are too long and technical for many people to use, and training for agency officials is inadequate at times. Participants felt that critical elements of efficient NEPA implementation included the extent to which an agency integrates NEPA's goals into its internal planning processes at an early stage and provides information to the public.

The study found that the extent to which the public is involved in the decision-making process also influences the potential for litigation. The study also found that some states, citizen groups, and businesses believe that certain EAs are prepared to avoid public involvement (i.e., because public meetings are not always required for EAs). The preparation of an EA, rather than an EIS, is reportedly the most common source of conflict and litigation under NEPA.[91] The study further found that nongovernmental organizations (NGOs) and citizens viewed the NEPA process as a one-way communication process, skeptical that their input was effectively incorporated into agency decision making and hypothesizing that their involvement was often solicited after decisions regarding actions and alternatives have already been made. Citizens also reported being frustrated when they were treated as adversaries rather than welcome participants in the NEPA process. Citizens reported that they often felt overwhelmed by the resources available to project proponents and agencies. As a consequence, litigation may be seen as the only means to affect environmental decisions significantly.[92]

In 2002, a comprehensive study of the NEPA process was conducted by CEQ's NEPA Task Force. CEQ formed the Task Force to review NEPA implementation practices and procedures and to determine opportunities to improve and modernize the process. The Task Force interviewed federal agencies; reviewed public comments, literature, and case studies; and spoke with individuals and representatives from state and local governments, tribes, and interest groups. In 2003, the Task Force released a report of its findings and recommendations.[93]

In compiling its research, the Task Force received more than 739 stakeholder comments. Those comments reflected current issues and challenges to NEPA implementation. With regard to delays in and the effectiveness of the NEPA process, a large percentage of comments were directed at factors related to NEPA analysis and documentation requirements and to the role and effects of litigation. According to CEQ, many respondents expressed a belief that the general requirement to provide adequate analysis had been taken to an extreme; that documents had become too time-consuming and costly to produce; and that the resultant "analysis paralysis" forestalled appropriate management of public lands and ultimately left the public distrustful and disengaged. The stakeholders felt this was brought on by vague requirements that were open to considerable interpretation and, therefore, an easy target for litigation. Because the requirements were vague, those commenters further felt that agencies were not sure how much analysis would be considered adequate by the courts, resulting in pressure to produce more.

In contrast, other respondents felt the "analysis paralysis" scenario was a misnomer. These respondents believed that agencies often predetermine the outcome of the planning process, that they often fail to consider other reasonable alternatives, and that the analysis agencies provide is often inadequate to support the management plan they propose. These commenters felt that the environmental effects of proposed actions are often inadequately considered, particularly the cumulative effects; that agencies rely on inadequate or outdated data; and that agency research is not held to the same rigorous standards as research in other fields, particularly in terms of scientific reference and peer review. Moreover, they felt that agencies are sometimes intent on following a predetermined course of action and ignore concerns submitted by the public. With regard to the role of litigation, a number of respondents felt that litigation only results when agencies do not comply with NEPA requirements. Some felt that it is only through litigation that concerned parties can get agencies to recognize their concerns and give serious attention to the environmental effects of their proposed actions.

One issue discussed in the Task Force report was challenges faced by agencies with regard to budget, training, and staffing constraints. This issue is discussed in more depth in a report, cited by the Task Force, that was prepared by the Natural Resources Council (an environmental conservation organization). That report surveyed 12 federal agencies to determine how they implemented the NEPA process.[94] Included in the report was a finding that, due to budget and staff constraints, most agencies' NEPA offices lack an ongoing national tracking system to monitor the numbers and types of NEPA documents that their agency is preparing or has completed. Also, the report found that agencies were unable to document their NEPA workload, calculate average preparation times or costs, show trends in these factors over time, or respond objectively to assertions that excessive time or money is being spent on complying with NEPA's requirements. The absence of such information, the report asserted, leaves agencies in a weak position to respond factually to or critically evaluate administrative or legislative proposals to "streamline" the NEPA process (see discussion, below).

Efforts to Streamline the NEPA Process

Some Members of Congress have expressed concerns that project delays are the result of inefficient interagency coordination required for large, complex projects. As a result, in the 108th Congress, several laws were enacted that included provisions intended to streamline the NEPA process. Although not defined in any of the legislation, the term "streamlining" was broadly used to describe legislative or administrative procedures intended to expedite the NEPA process.[95] It usually refers to a process or procedures to better coordinate federal, state, tribal, or local agency action, when compliance with multiple environmental laws, regulations, or executive orders is required.

In 2008, most agencies filed fewer than 10 EISs.[96] Just over 63% of all EISs were filed by seven agencies—USDA-Forest Service (124); DOT's Federal Highway Administration (64) and Federal Transit Administration (21); the Department of the Interior's Bureau of Land Management (48) and National Park Service (25); the Department of Defense's Army Corps of Engineers (42), and the Department of Energy's Federal Energy Regulatory Commission (19). It may not be surprising, then, that many streamlining activities involve actions sponsored by those agencies. For example, what follows are bills enacted since the 108th Congress[97] and selected types of projects for which streamlining provisions have been included:

- The Healthy Forests Restoration Act of 2003 (P.L. 108-148): "hazardous fuel reduction" projects on federal land.
- Vision 100-Century of Aviation Reauthorization Act (P.L. 108-176): airport capacity enhancement projects at congested airports.

- The Safe, Accountable, Flexible, and Efficient Transportation Equity Act of 2005: A Legacy for Users (P.L. 108-59): construction of or modifications to surface transportation projects.
- The Energy Policy Act of 2005 (P.L. 108-58): various energy development projects, such as oil and gas leasing and permitting on federal land, and the designation of energy facility rights-of-way and corridors on federal lands.
- The Water Resources Development Act (WRDA) of 2007 (P.L. 110-114): water resources projects undertaken by the Army Corps of Engineers.

Streamlining provisions are unique to the class of projects at issue. However, most include some or all of the following elements:

- The designation of specific projects as categorical exclusions.
- The designation of a specific agency as the "lead agency" for all classes of certain actions (e.g., delegation of DOT as the lead agency for all highway or transit projects requiring review under NEPA).
- Direction to the lead agency to develop a "coordinated environmental review" process to ensure early coordination and cooperation among federal, state, tribal, and local agencies required to participate in a project.
- Delegation of specific authority to the lead agency, such as the authority to establish deadlines for cooperating agencies, specify a project's "purpose and need," or specify project alternatives.
- Delegation of certain federal authority to state or local agencies (e.g., the authority to determine whether certain classes of projects may be categorically excluded from environmental requirements).
- Direction to the lead agency to develop dispute resolution procedures if agencies reach an impasse in the NEPA process.

Streamlining proposals have generated a great deal of controversy among interested stakeholders (e.g., agency representatives, industry groups, and environmental organizations). Most stakeholders agree that the process for complying with environmental requirements applicable to complex federal projects can be implemented more efficiently. How that should be done and the degree to which it is necessary have been the subject of considerable debate. Some stakeholders, such as industry representatives who would like to see projects implemented more quickly, argue that the authority of lead agencies must be strengthened to reduce delays caused by disagreements among agencies. They also contend that lead agencies should have the authority to set and enforce deadlines with regard to the cooperating agency decision-making process. Environmental groups are concerned that by speeding up the compliance process and strengthening lead agency authority, concerns of the public or cooperating agencies will be minimized or ignored, in effect rubber stamping lead agency decisions. Further, some environmental groups contend that "streamlining" is a thinly veiled attempt at weakening environmental protection and reducing public participation in the federal decision-making process.

CONCLUSION

NEPA is a procedural statute that, along with CEQ and individual agencies' regulations, specifies procedures that must be followed in the federal decision-making process. It imposes no requirement other than to require agencies to consider the environmental impacts of their actions before proceeding with them and to involve the public in that process. It does not dictate what the decision must be. More specifically, it does not require the agency to select the least environmentally harmful alternative or to elevate environmental concerns above others.

The role the courts have played in NEPA's implementation is arguably more pronounced compared to many other environmental laws because of several unique factors. These include the initial lack of binding regulations applicable to the EIS preparation process, the absence of an agency authorized to enforce its requirements, and NEPA's requirement to involve the public in the decision-making process. With regard to the latter, when members of the public oppose a project or feel that their opinions are not given sufficient weight, their involvement may result in turning to the courts to halt the project until their concerns are addressed. During the past 35 years, interested stakeholders have challenged the adequacy of NEPA documentation and agency compliance with NEPA in court and, in some instances, used NEPA litigation to try to halt or slow projects to which they were opposed. As a result, the progress of some federal projects was slowed. However, particularly in the past 10-15 years, the number of projects affected by NEPArelated litigation is very small. Also, unlike other environmental laws, NEPA itself cannot stop a project altogether. This does not mean that, during the course of a NEPA-related lawsuit, an agency may not decide to abandon a given project or project alternative.

As a policy statute, NEPA supplements other statutes. Consequently, agencies often are required to comply with provisions of other state, tribal, and federal environmental requirements before they can proceed with a given action. This requirement can lead to confusion when procedures to comply with other laws are integrated with NEPA compliance, and it can give the impression that NEPA alone is responsible for the time it takes to obtain the appropriate authorization or approval for a federal project.

Although stakeholders disagree about the extent to which NEPA currently halts or delays federal actions, few disagree that agencies can improve their methods of NEPA compliance. Many elements of recent legislative proposals intended to streamline NEPA compliance already exist in CEQ's regulations. Those include integrating NEPA early in the planning process, integrating NEPA requirements with other environmental requirements, eliminating duplication with state and local procedures, swiftly addressing disputes with other agencies, and establishing appropriate time limits on the EIS process. Debate is likely to continue with regard to if or to what degree further streamlining may be accomplished.

SELECTED REFERENCES

Bear, Dinah. "NEPA at 19: A Primer on an 'Old' Law with Solutions to New Problems," 19 *Environmental Law Reporter* 10060, February 1989.

Caldwell, Lynton. *The National Environmental Policy Act: An Agenda for the Future.* Bloomington, IN: Indiana University Press, 1998.

Council on Environmental Quality. "Forty Most Asked Questions Concerning CEQ's National Environmental Policy Act Regulations," 46 *Federal Register* 18026, March 23, 1981.

Council on Environmental Quality. "Memorandum for General Counsels, NEPA Liaisons and Participants in Scoping," April 20, 1981.

Council on Environmental Quality. "The National Environmental Policy Act: A Study of Its Effectiveness After Twenty-five Years," January 1997.

Council on Environmental Quality's NEPA Task Force. "Report to the Council on Environmental Quality: Modernizing NEPA Implementation," September 2003.

Smythe, Robert and Isber, Caroline. "NEPA in the Agencies: 2002, A Report to the Natural Resources Council of America," October 2002.

End Notes

[1] CEQ is charged with providing oversight and guidance to agencies with regard to EIS preparation. The Environmental Protection Agency (EPA) is required to review and comment publicly on the environmental impacts of proposed federal activities, including those for which an EIS is prepared. EPA is also the official recipient of all EISs prepared by federal agencies. However, neither agency has enforcement authority with regard to an agency's environmental review requirements.

[2] S.Rept. 91-296, 91st Cong., 1st sess., July 9, 1969, pp. 11-12, and Lynton Caldwell, *The National Environmental Policy Act: An Agenda for the Future*, Indiana University Press, 1998, pp. 26-27.

[3] Statement by Governor Rockefeller in *Joint House-Senate Colloquium to Discuss a National Policy for the Environment, Hearing before the Committee on Interior and Insular Affairs, United States Senate, and the Committee on Science and Astronautics, U.S. House of Representatives*, July 17, 1968, p. 5.

[4] "Congressional White Paper on a National Policy for the Environment," issued jointly by the Committee on Interior and Insular Affairs, United States Senate, and the Committee on Science and Astronautics, U.S. House of Representatives, summarizing key points raised in the dialog, October 1968, Committee Print, p. 9.

[5] S. 1075 was introduced by Senator Jackson on February 18, 1969. H.R. 12549 was introduced by Representative John Dingell, and others, on July 1, 1969.

[6] Senator Jackson's remarks regarding "S. 1075—Introduction of Bill to Establish a National Strategy for the Management of the Human Environment," vol. 115, *Congressional Record*, p. S1780, February 18, 1969 (reprinted in *National Environmental Policy Act of 1969: Legislative History, Senate bill 1075, Public law 91-190, 91st Congress, 1st session*, James D. Nuse, compiler, for the U.S. Atomic Energy Commission, September 1970).

[7] Although several similar bills were introduced in the Senate that session, those that ultimately received consideration by the Committee were S. 1075, S. 237 (which included a declaration of environmental policy) and S. 1752.

[8] Dr. Caldwell is a Professor of Public and Environmental Affairs at Indiana University and was a staff consultant to the Senate Committee on Interior and Insular Affairs on a National Policy for the Environment, 1968-1970.

[9] *Hearing before the Committee on Interior and Insular Affairs, United States Senate, 91st Congress, 1st Session, on S. 1075, S. 237, and S. 1752*, April 16, 1969, p. 116.

[10] Ibid., p. 117.

[11] S. 1075 Amendments, referred to Senate Committee on Interior and Insular Affairs, May 29, 1969.

[12] Section 102(C), S. 1075 Amendments, May 29, 1969.

[13] S.Rept. 91-296, p. 14.

[14] NEPA was amended by P.L. 94-52, July 3, 1975, regarding how CEQ may spend appropriated funds; P.L. 94-83, August 9, 1975, specifying parameters under which states may prepare an EIS; and P.L. 97-258, § 4(b), September 13, 1982, regarding budget and accounting procedures.

[15] 42 U.S.C. § 7609.

[16] 42 U.S.C. § 4331.

[17] 42 U.S.C. § 4331(b).

[18] 42 U.S.C. § 4332(2)(C).

[19] Ibid.

[20] 42 U.S.C. § 4332(2)(E).

[21] 42 U.S.C. § 4333.

[22] 42 U.S.C. § 4335.

[23] 42 U.S.C. § 4341. The 1997 Environmental Quality Report was the last one prepared by CEQ. The Federal Reports Elimination and Sunset Act of 1995 (P.L. 104-66) eliminated many congressionally mandated reports, including the annual CEQ Environmental Quality Report, unless explicitly requested by Congress.

[24] Lynton Caldwell, *The National Environmental Policy Act: An Agenda for the Future*, Indiana University Press, 1998, p. 2.

[25] For an analysis of legal issues, consult the American Law Division of CRS.

[26] *Baltimore Gas & Electric Co. v. Natural Resources Defense Council, Inc.*, 462 U.S. 87, 97, 100 (1983).

[27] *Robertson v. Methow Valley Citizens Council*, 490 U.S. 332 (1989).

[28] Ibid., at 351.

[29] Executive Order 11514, Protection and Enhancement of Environmental Quality, signed by President Nixon, March 5, 1970, 35 *Federal Register* 4247.

[30] H.Rept. 92-316, "Administration of the National Environmental Policy Act," June 1971.

[31] Dinah Bear (CEQ General Counsel), "NEPA at 19: A Primer on an 'Old' Law with Solutions to New Problems," 19 *Environmental Law Reporter* 10060, p. 10062, February 1989.

[32] Dinah Bear, "The National Environmental Policy Act: Its Origins and Evolutions," *Natural Resources & Environment*, vol. 10, no. 2, fall 1995, p. 70.

[33] Executive Order 11991, Relating to Protection and Enhancement of Environmental Quality, signed by President Carter, May 24, 1977, 42 *Federal Register* 26967.

[34] 43 *Federal Register* 55978, November 28, 1978; 40 C.F.R. §§ 1500-1508.

[35] Council on Environmental Quality, *Ninth Annual Report of the Council on Environmental Quality*, December 1978, pp. 396-399; and at 40 C.F.R. §§ 1500.4 and 1500.5.

[36] 40 C.F.R. § 1500.1.

[37] 40 C.F.R. § 1507.3.

[38] 40 C.F.R. § 1500.6.

[39] The "Forty Most Asked Questions" and other CEQ guidance are available at http://ceq.hss.doe.gov/nepa/regs/ guidance.html

[40] The guidance is available at http://www.whitehouse.gov/administration/eop/ceq/ initiatives/nepa.

[41] The CEQ regulations, at 40 C.F.R. § 1508.18, specify that the term "major" in the phrase "major federal action" reinforces but does not have meaning independent of "significantly." This discussion focuses on criteria used to determine whether the level of federal agency involvement in an action is such that the action is federalized and, when it is, the categories of action subject to NEPA. Criteria used to determine if environmental impacts are significant are discussed in "Determining the Significance of a Federal Action" section, below.

[42] 40 C.F.R. § 1508.18(a).

[43] 40 C.F.R. § 1508.18(b)(4). Further, the term "federal agency" is defined as all agencies of the federal government but does not mean the Congress, the Judiciary, or the President (40 C.F.R. § 1508.12).

[44] Sixteen states, the District of Columbia, Puerto Rico, and some municipalities and tribal governments have enacted their own environmental quality acts, sometimes referred to as "little NEPAs." For a list of the 16 states and corresponding citations to relevant state law,

see the State Environmental Resource Center (SERC) web page regarding State Environmental Quality Acts at http://www.serconline.org/SEQA/stateactivity.html.

[45] Determining the level of federal involvement that would federalize a private project is becoming more relevant as state and local governments are increasingly turning to public-private partnerships to fill gaps in federal funding for needed projects (e.g., surface transportation projects, school maintenance, and construction projects).

[46] The tiering process is discussed at 40 C.F.R. § 1502.20; also, see answers to questions 24a-c regarding EISs required on policies, plans, or programs in CEQ's "Forty Most Asked Questions."

[47] 40 C.F.R. § 1508.17.

[48] 40 C.F.R. § 1508.27.

[49] 40 C.F.R. § 1508.27(a).

[50] 40 C.F.R. § 1508.27(b).

[51] CEQ's "Number of EIS filed with EPA by Federal Agencies," available at http://ceq.hss.doe.gov/nepa/nepanet.htm.

[52] See "FHWA Projects by Class of Action," on FHWA's "Streamlining/Stewardship" Website, available at http://environment.fhwa.dot.gov/strmlng/projectgraphs.asp.

[53] 40 C.F.R. § 1508.22.

[54] 40 C.F.R. § 1501.7.

[55] 40 C.F.R. § 1502.9.

[56] 40 C.F.R. § 1502.14.

[57] See answer to question 2a regarding Alternatives Outside the Capability of Applicant or Jurisdiction of Agency in CEQ's guidance document "Forty Most Asked Questions."

[58] Ibid., answer to question 1b regarding how many alternatives have to be discussed when there is an infinite number of possible alternatives.

[59] 40 C.F.R. § 1505.2.

[60] Copies of this documentation are generally available from the lead agency. Often, particularly for EISs, NEPA documentation is available on an agency's website.

[61] 40 C.F.R. § 1508.9(a).

[62] 40 C.F.R. § 1508.9(b).

[63] 40 C.F.R. § 1501.4(e)(1).

[64] 40 C.F.R. § 1507.3.

[65] 23 C.F.R. § 771.117.

[66] 40 C.F.R. § 1508.4.

[67] The guidance is available online at http://www.whitehouse.gov/administration/eop/ceq/initiatives/nepa.

[68] 40 C.F.R. § 1508.16.

[69] A project applicant may be any public or private entity initiating a "federal" action. For example, a state department of transportation may initiate the NEPA process for a federally funded surface transportation project. A project applicant may also be a private entity requesting a federal permit. For example, a livestock owner applying for a permit to graze cattle on federal land may also initiate the NEPA process. The lead agency in such a case would be the federal agency (e.g., the U.S. Department of Agriculture's Forest Service) issuing the permit.

[70] 40 C.F.R. § 1508.5.

[71] Text of Secretary Mineta's May 6, 2003 letter, and Chairman Connaughton's May 12, 2003 response, are available at http://www.environment.fhwa.dot.gov/guidebook/ Ginterim.asp.

[72] 42 U.S.C. § 4332(2)(C).

[73] 40 C.F.R. § 1503.

[74] 40 C.F.R. § 1503.1.

[75] 40 C.F.R. §1503.4.

[76] NEPA documentation is required of EPA for research and development activities, facility construction, wastewater treatment plant construction under the Clean Water Act, EPA-issued National Pollutant Discharge Elimination System (NPDES) permits, and certain projects funded through EPA annual appropriations acts. For more information about EPA's requirements with regard to NEPA compliance, see http://www.epa.gov/compliance/nepa/epacompliance/ index.html.

[77] See an explanation of EPA's "Environmental Impact Statement (EIS) Rating System Criteria" at http://www.epa.gov/compliance/nepa/comments/ratings.html.

[78] 40 C.F.R. §1504.1.

[79] 40 C.F.R. §1504.2.

[80] Although the MOA is not readily available, reference to it and the allocation of duties between EPA and CEQ is discussed in a March 7, 1989, *Federal Register* notice available on EPA's Compliance and Enforcement website under "EIS Filing System," at http://www.epa.gov/compliance/resources/policies/nepa/index.html.

[81] 40 C.F.R. § 1500.2(d).

[82] 40 C.F.R. § 1506.6.

[83] Council on Environmental Quality, *The National Environmental Policy Act: A Study of Its Effectiveness After Twenty-five Years*, January 1997, p. ix.

[84] 40 C.F.R. § 1502.25.

[85] See, Council on Environmental Quality "2005 Litigation Survey," available at http://ceq.eh.doe.gov/nepa/ nepanet.htm.

[86] See Council on Environmental Quality, *Environmental Quality: 25th Anniversary Report* (the CEQ 1994-95 Annual Report) 1996, p. 51, available at http://ceq.eh.doe.gov/nepa/reports/reports.htm, and Litigation Surveys for 2001 through 2005, available at http://ceq.eh.doe.gov/nepa/nepanet.htm.

[87] Dinah Bear, "NEPA at 19," 19 *ELR* 10062.

[88] Council on Environmental Quality, "NEPA Study of Effectiveness After Twenty-five Years," p. iii.

[89] The survey, *Fast Tracking NEPA Documents—Tools to Overcome Schedule Delays,* was conducted in 2003 and 2004 by Tetra Tech, Inc., for presentation at the 30th National Defense Industrial Association Environmental and Energy Symposium and Exhibition. The survey includes responses from agency staff identified as NEPA project managers, NEPA resource authors, agency NEPA officers or reviews, or non-NEPA professionals, such as engineers. Results are available at http://www.dtic.mil/ndia/2004enviro/, under April 7, 2004, Session Nine.

[90] Council on Environmental Quality, "NEPA Study of Effectiveness After Twenty-five Years."

[91] Ibid., p. 19.

[92] Ibid., p. 18.

[93] "The NEPA Task Force Report to the Council on Environmental Quality: Modernizing NEPA Implementation," September 2003, and comments documented by the Task Force are available at http://ceq.eh.doe.gov/ntf/index.html.

[94] Robert Smythe and Caroline Isber, "NEPA in the Agencies: 2002, A Report to the Natural Resources Council of America," October 2002, available at http://www.naturalresources council.org/ewebeditpro/items/O89F2656.pdf. Many of the findings regarding agency

implementation of NEPA and recommendations for change were directed to the CEQ NEPA Task Force.

[95] The term "streamlining" has also been used to refer to administrative or legislative actions intended to expedite the process of complying with other environmental requirements, such as permitting.

[96] See the Council on Environmental Quality's "Number of EIS filed with EPA by Federal Agencies" for 2008, available at http://ceq.hss.doe.gov/nepa/ Calendar_Year_2008_ Filed_EISs.pdf.

[97] Since 2007, there have been various legislative proposals with provisions intended to expedite the NEPA process or waive NEPA requirements for certain classes of federal action, but few legislative proposals have progressed beyond their introduction.

In: The National Environmental Policy Act ISBN: 978-1-61324-876-8
Editors: M. Drubich and A. Moser

Chapter 2

OVERVIEW OF NATIONAL ENVIRONMENTAL POLICY ACT (NEPA) REQUIREMENTS*

Kristina Alexander

SUMMARY

The National Environmental Policy Act (NEPA) establishes environmental policies that apply to the federal government, but it is best known for imposing environmental review procedures on federal agency actions. NEPA requires agencies to review the potential environmental impacts of their projects, recording their review in a publicly available document. There are three types of environmental documents, based on the type of review: a categorical exclusion (CE), an environmental assessment (EA), and an environmental impact statement (EIS). Tiering, programmatic, and supplemental documents are subsets of these original documents.

The act dictates procedure, not results. NEPA requires agencies to follow the process so that their decisions are informed. Agencies are required to take a "hard look" at the environmental impacts, not to meet set environmental standards nor to choose the project with the least environmental consequence. This report is intended as a general introduction to NEPA, analyzing what NEPA requires for those three types of environmental documents.

---

* This is an edited, reformatted and augmented version of Congressional Research Service Report RS20621, dated January 12, 2011.

GENERAL

The National Environmental Policy Act of 1969 (NEPA)[1] establishes environmental policies that apply to the federal government, but it is best known for imposing environmental review procedures on federal agency actions. Except as otherwise provided by Congress, the act applies to all federal agency actions,[2] including those that intersect with private activities, such as through a federal permit or funding, although its requirements vary depending on the nature of the action involved.[3]

NEPA created the Council on Environmental Quality (CEQ), which promulgated regulations implementing the act.[4] These regulations apply to all agencies, and the U.S. Supreme Court has held that they are entitled to deference.[5] The NEPA regulations emphasize communicating with the public, reducing delays of federal projects, and making better decisions.[6] Agencies are to integrate NEPA reviews with other agency planning and review processes,[7] and coordinate with other federal agencies and with similar state processes when appropriate.[8] Each agency is expected to elaborate on how to comply with NEPA in the context of its own duties. Agencies generally issue regulations specifying their NEPA review process.[9] When more than one federal agency is involved in an action, the regulations provide for the responsibilities of a "lead agency"[10] and "cooperating agencies."[11]

NEPA establishes goals for agency actions. It sets a national environmental policy that the federal government should "use all practicable means" to improve and coordinate federal actions to assure "safe, healthful, productive, and aesthetically and culturally pleasing surroundings" and to "attain the widest range of beneficial uses of the environment without degradation, risk to health or safety, or other undesirable and unintended consequences."[12]

In addition to this policy, NEPA requires certain practices. The most significant of these practices are the requirements in Section 102(2), which is where the requirement for preparing an environmental impact statement (EIS) is found. It states that "for proposals for legislation and other major Federal actions significantly affecting the quality of the human environment" the agency must prepare a detailed environmental review discussing the following:

I. the environmental impact of the proposed action,
II. any adverse environmental effects which cannot be avoided should the proposal be implemented,

III. alternatives to the proposed action,
IV. the relationship between local short-term uses of man's environment and the maintenance and enhancement of long-term productivity, and
V. any irreversible and irretrievable commitments of resources which would be involved in the proposed action should it be implemented.[13]

These environmental reviews are to be conducted early in the decision-making process so that they can contribute to the process and not be used to "rationalize or justify decisions already made."[14]

NEPA DOCUMENTS

An environmental review under NEPA can take one of three forms: a categorical exclusion (CE), an environmental assessment (EA), or an environmental impact statement (EIS). For each of these, the CEQ regulations require consideration of the possible environmental effects of an agency action. The extent of that analysis depends on the circumstances and the likely degree of environmental impacts.

CEs are at one end of the spectrum. CEs are used for categories of actions that have been determined not to have a significant effect on the human environment, individually or cumulatively, and therefore, do not require further analysis.[15] Typically, CEs are used for minor actions that an agency does repeatedly and knows will have no or only minor environmental effects. Agencies publish lists of activities that are CEs. When a project arises that falls into a category on the list, a CE may be invoked.[16] However, if an "extraordinary circumstance" (such as wetlands or a threatened or endangered species) is present, a CE may not be used.[17] Even though a CE represents a determination that an environmental review is not necessary, the determination still must be documented and cannot be made after a project has begun.[18]

In the middle of the spectrum are EAs. EAs are conducted to determine whether an EIS is needed or a finding of no significant impact (FONSI) is appropriate.[19] EAs are intended to be concise, but are also required to consider the need for the project, the environmental impacts of the project and its alternatives, alternatives required by section 102(2)(E), and a list of the agencies and persons consulted.[20] They undergo a draft version before the final EA is issued. The CEQ regulations do not require public comment for draft EAs, although some agencies do circulate those documents for comment.

At the other end of the spectrum are full reviews for major federal actions that may significantly affect the environment—EISs.[21] These documents are the most complex and have drawn the most scrutiny. The thoroughness with which an agency must study environmental effects and consider alternatives is greatest if an EIS is required.[22] An EIS must discuss an adequate range of proposed alternatives, and the direct, indirect, and cumulative effects or impacts of each. The required documents can be voluminous and may take years to produce. Under NEPA, different versions of EISs are produced: DEIS (draft EIS), which is circulated for comment; FEIS (final EIS); SEIS (supplemental EIS); and DSEIS (draft supplemental EIS), also circulated for comment.[23] Additionally, the type of an EIS can vary depending on the nature of the project. A site-specific EIS is prepared for one project. A programmatic EIS is prepared for a series of related projects. A generic EIS is prepared for a type of project that will be repeated. A tiered analysis may be used for interrelated projects, such as following a programmatic EIS with an EA for a specific part of the project, incorporating by reference the analyses from the earlier document.[24]

Even though both EISs and EAs require reviews of environmental impacts, there is a difference between these documents. An EA review is intended to be briefer. Its purpose is to find whether there are significant environmental impacts. If there are significant impacts, an EIS is developed to analyze those impacts in detail. Both documents must review not only the impacts of the planned project, but the impacts of the alternatives. However, there is a distinction between the review within an EA and that of an EIS. The regulations require an EA to "include brief discussions" on the impacts of the alternatives. In contrast, an EIS's alternatives analysis is referred to as the "heart of the environmental impact statement."25 The document must "devote substantial treatment to each alternative."[26] A successful EIS will clearly illustrate how the agency made its decision to choose that alternative. NEPA does not require the agency to choose the most environmentally preferable alternative.

CEs, by their nature, are extremely brief. Agencies have been encouraged to create more CEs[27] because CEs are perceived to be less of a burden on agency resources than EAs or EISs.[28] For example, the CEQ issued draft guidance in 2006 encouraging agencies to designate more CEs.[29] Draft guidance in 2010 noted that inappropriate use of CEs could "thwart the purposes of NEPA, compromising the quality and transparency of agency decisionmaking."[30] In addition to administratively created CEs, Congress has established some legislatively. Three significant pieces of federal legislation directed use of CEs: the Energy Policy Act of 2005 (for projects disturbing fewer than five acres); the Safe, Accountable, Flexible, and Efficient Transportation Equity Act: A Legacy for Users of 2005 (for activities using intelligent transportation infrastructure and systems); and the Healthy Forests Restoration Act of 2003 (P.L. 108-148) (for hazardous fuel reduction projects on federal land). Additionally, Congress can find that CEs are not appropriate and end their use. Following the 2010 Gulf Oil Spill, the 111th Congress considered legislation to end CEs for certain offshore drilling practices.[31] Agencies update their lists of CEs regularly.

PUBLIC PARTICIPATION

NEPA prescribes federal agency review but does not specify public involvement in the reviews. However, the NEPA regulations discuss and require public participation. The public participation aspects of the NEPA process are regarded by many as a valuable aspect of the law. Agencies must

> provide public notice of NEPA-related hearings, public meetings, and the availability of environmental documents so as to inform those persons and agencies who may be interested or affected.[32]

Public input is encouraged early in the process. CEQ recommends that a scoping process be conducted for EISs and that public comment be included in the scoping process.[33] The scoping process is intended to bring interested parties together, identify the significant issues at an early stage to narrow the environmental review, decide whether to prepare an EA or an EIS, and set time limits for the review.[34] The CEQ regulations specify levels of notice to the public depending on whether an action is of national or local interest, and state that in all cases the agency must mail notice to those who requested it regarding a particular action.[35] Also, draft EISs must be made available for public comment.[36] Those comments must be responded to by the lead agency in the final version of the document.[37] Court challenges to EISs must wait until there is a final agency action. EISs are final only upon the issuance of the Record of Decision (ROD). An EA becomes final when a FONSI or a notice of intent to prepare an EIS is issued.

Public hearings are appropriate for any type of environmental document when there is substantial controversy regarding a project or substantial interest in having a hearing.[38]

No public involvement in the preparation of an EA or in making a categorical exclusion determination is specified by the CEQ regulations, although agency-specific regulations may require some public notice for these determinations.[39] Notably, the CEQ regulations for these documents do not require drafts to be issued to the public.

NEPA LITIGATION

Because NEPA does not provide a right of action, suit is brought under the Administrative Procedure Act (5 U.S.C. §§ 706 *et seq.*). Like other APA cases, the court reviews the administrative record to see if the agency acted arbitrarily or capriciously. NEPA litigation is complicated, and just about every word in "major Federal actions significantly affecting the quality of the human environment" has been disputed, scrutinized, and defined by the courts. Some issues have been resolved. For example, there is consensus that a court's review of a NEPA document is to ensure that the agency took a "hard look" at the environmental impacts of an action.[40] Also, the courts agree that NEPA "does not mandate particular results, but simply prescribes the necessary process."[41] Thus, "NEPA merely prohibits uninformed—rather than unwise—agency action."[42] Most court cases relate to when an EIS needs to be prepared and to the adequacy of EIS coverage.[43]

For EAs, the contention generally is whether that document demonstrates that no EIS is required. Sometimes this can take the form of a claim that the agency had decided that an EIS was not required *before* preparing the EA, which is against the express purpose of using the document to inform decision making. The courts review EAs to determine whether the agency considered the alternatives and the impacts from those alternatives.

CEs are different. They are intended to allow agencies to find that a project has no significant effects on the environment without conducting a review, but the agency must still assemble a record to support its decision. Disputes regarding CEs center on whether the category was proper in the first place (such as the existence of extraordinary circumstances) or whether the designation of a CE for a particular project was appropriate. One factor in determining whether a CE was appropriate is whether the action normally requires an EA or an EIS.[44]

End Notes

[1] Act of January 1, 1970, P.L. 91-190, 83 Stat. 852; 42 U.S.C. §§ 4321 *et seq.* For a more general discussion of NEPA, see CRS Report RL33152, *The National Environmental Policy Act (NEPA): Background and Implementation*, by Linda Luther.

[2] 42 U.S.C. § 4332 ("The Congress authorizes and directs that, to the fullest extent possible ... all *agencies* of the Federal Government shall [do the following]") (emphasis added). 40 C.F.R. § 1508.12 defines "Federal agency" as not meaning the Congress, the Judiciary, or the President. Attention focuses on this regulation from time to time, e.g., when the President established several national monuments without a NEPA review.

[3] There are also several judicially created exemptions for when NEPA does not apply, although they are the exception. Some examples include appropriations, where the statute provides no discretion, and where the statutory deadline is too short.

[4] 40 C.F.R. §§ 1500 *et seq.* CEQ is within the Executive Office of the President with three members appointed by the President. 42 U.S.C. § 4342. It does not have the responsibility for enforcing NEPA requirements. Each agency oversees its own compliance with the act.

[5] Andrus v. Sierra Club, 442 U.S. 347, 358 (1979).

[6] 40 C.F.R. §§ 1500.1 – 1500.5.

[7] 40 C.F.R. § 1500.2(c).

[8] 40 C.F.R. § 1506.2.

[9] See, e.g., Federal Highway Administration (FHWA) regulations at 23 C.F.R. Part 771; United States Department of Agriculture (USDA) regulations at 7 C.F.R. Part 1b; Housing and Urban Development (HUD) regulations at 24 C.F.R. Part 50. Some agency NEPA regulations are online at http://ceq.hss.doe.gov/NEPA/regs/agency/agencies.cfm.

[10] 40 C.F.R. § 1501.5.

[11] 40 C.F.R. § 1508.5.

[12] NEPA § 101(b); 42 U.S.C. § 4331(b).

[13] 42 U.S.C. § 4332(2)(C).

[14] 40 C.F.R. § 1502.5. See also 40 C.F.R. § 1501.2.

[15] 40 C.F.R. § 1508.4.

[16] For example, FHWA has determined these actions are suitable for CEs: landscaping, construction of bike and pedestrian lanes, improvements to existing truck stations (23 C.F.R. § 771.117). The Forest Service lists these CEs: closing an area during a period of extreme fire danger, adjusting special use or recreation fees, approving a Surface Use Plan for oil and natural gas exploration in a new field without herbicides (FSH 1909.15_31). CEs in HUD include inspections and testing of properties for hazards or defects, purchase of tools, and approval of foreclosure sale of HUD-held mortgages (24 C.F.R. § 50.19).

[17] 40 C.F.R. § 1508.4. Some examples of "extraordinary circumstances" include highly controversial issues (see *Fund for Animals v. Babbitt*, 89 F.3d 128, 133 (2d Cir. 1996)); actions that cumulatively may have significant effects (see *Citizens for Better Forestry v. USDA*, 481 F. Supp. 2d 1059, 1089 (N.D. Cal. 2007)); and actions that may have significant effects, such as by affecting an endangered species or an archeological site.

[18] See Wilderness Watch and Public Employees for Environmental Responsibility v. Mainella, 375 F.3d 1085 (11th Cir. 2004) (NEPA was violated where record indicated that a categorical exclusion was invoked only after the action had been taken); Anacostia Watershed Soc'y v. Babbitt, 871 F. Supp. 475, 481 (D.D.C. 1994); Fund for Animals, Inc. v. Espy, 814 F. Supp. 142, 150-51 (D.D.C. 1993) (a categorical exclusion must be invoked prior to the agency action).

[19] 40 C.F.R. § 1508.9(a)(1). An EA that results in a FONSI is referred to as an EA-FONSI.

[20] 40 C.F.R. § 1508.9(b).

[21] 40 C.F.R. § 1501.4, and parts 1502 and 1503.

[22] Mt. Lookout - Mt. Nebo Property Protection Ass'n v. FERC, 143 F.3d 165 (4th Cir. 1998).

[23] See 40 C.F.R. § 1502.9 for a description of these different types of documents.

[24] 40 C.F.R. § 1502.20.

[26] 40 C.F.R. § 1502.14(b).

[27] See 48 Fed. Reg. 34263, 34265 (July 28, 1983) (criticizing excessive use of EAs, and encouraging agencies to draft broadly defined criteria for CEs).

[28] "Establishing, Revising, and Using Categorical Exclusions under the National Environmental Policy Act," 70 Fed. Reg. 54816, 54817 (September 19, 2006).

[29] 71 Fed. Reg. 54816 (September 19, 2006). No final guidance has been issued as of April 9, 2009.

[30] Chair, Council on Environmental Quality, "Memorandum for Heads of Federal Departments and Agencies" (February 18, 2010). Available online at http://ceq.hss.doe.gov/nepa/regs/Categorical_Exclusion_Draft_NEPA_Guidance_FINAL_02182010.pdf.

[31] See, e.g., H.R. 3534 (the CLEAR Act), H.R. 5506, and H.R. 5572 from the 111th Congress.

[32] 40 C.F.R. § 1506.6(b).

[33] 48 Fed. Reg. 34263 (July 28, 1983). See also Forest Service Handbook, 1909.15, ch. 30, requiring an abbreviated scoping process for categorical exclusions.

[34] 40 C.F.R. § 1501.7.

[35] 40 C.F.R. § 1506.6(b).

[36] 40 C.F.R. § 1503.1.

[37] 40 C.F.R. § 1503.4.

[38] 40 C.F.R. § 1506.6(c)(1).

[39] For example, the FHWA regulations require that EAs are made available for public review, but not comment. 23 C.F.R. § 771.119(d).

[40] Natural Resources Defense Council v. Morton, 458 F.2d 827, 838 (D.C. Cir. 1972).

[41] Robertson v. Methow Valley Citizens Council, 490 U.S. 332, 350 (1989).

[42] Id. at 351.

[43] See THE NEPA LITIGATION GUIDE, Karin Sheldon and Mark Squillace, eds. (American Bar Association, 1998).

In: The National Environmental Policy Act ISBN: 978-1-61324-876-8
Editors: M. Drubich and A. Moser

Chapter 3

IMPLEMENTING THE NATIONAL ENVIRONMENTAL POLICY ACT (NEPA) FOR DISASTER RESPONSE, RECOVERY, AND MITIGATION PROJECTS*

Linda Luther

SUMMARY

In the aftermath of a major disaster, communities may need to rebuild, replace, or possibly even relocate a multitude of structures. When recovery activities take place on such a potentially large scale, compliance with any of a number of local, state, and federal laws or regulations may apply. For example, when older buildings must be repaired or demolished, provisions of the National Historic Preservation Act (NHPA) may need to be considered. If rebuilding will take place in a floodplain, provisions of Executive Order 11988 on Floodplain Management may apply.

When federal agencies make decisions, such as funding applicant-proposed actions, the National Environmental Policy Act of 1969 (NEPA, 42 U.S.C. § 4321 *et seq.*) applies. For example, when federal funding is provided for disaster-related activities, applicants for those funds may be required to assess the environmental impacts of their proposed action. As

* This is an edited, reformatted and augmented version of Congressional Research Service Report RL34650, dated January 13, 2011.

commonly implemented, NEPA's environmental review requirements are used as a vehicle to identify *any other* environmental requirements that may apply to a project as well. This use of NEPA as an "umbrella" statute can lead to confusion. For example, before an applicant can commit or expend funds under the Department of Housing and Urban Development's (HUD's) Community Development Block Grant (CDBG) program, the applicant must complete an environmental review of the project. A required element of that review is the applicant's certification that compliance criteria applicable to historic preservation, floodplain management, endangered species, air quality, and farmland protection have been considered. This review is required not only to meet NEPA obligations, but also to ensure that the project being funded does not violate applicable environmental law. From the applicant's perspective, this may blur the distinction between what is required under NEPA and what is required under separate compliance requirements *identified* within the context of the NEPA process.

For many federal actions undertaken in response to emergencies or major disasters, NEPA's environmental review requirements are exempted under provisions of the Robert T. Stafford Disaster Relief and Emergency Assistance Act (the Stafford Act). (The Stafford Act does not, however, exempt such projects from other applicable environmental requirements.) In the past, some members of Congress have been interested in the NEPA process as it applies to disaster-related projects. This interest has been driven, in part, by federal grant applicants who have been confused about both their role in the NEPA process and what the law requires.

To address issues associated with the NEPA process, this report discusses NEPA as it applies to projects for which federal funding to recover from or prepare for a disaster has been requested by local, tribal, or state grant applicants. Specifically, the report provides an overview of the NEPA process as it applies to such projects, identifies the types of projects (categorized by federal funding source) likely to require environmental review, and delineates the types of projects for which no or minimal environmental review is required (i.e., those for which statutory or regulatory exemptions apply) and those likely to require more in-depth review.

INTRODUCTION

The federal government administers various programs to assist individuals and communities in responding to, recovering from, and preparing for disasters. For example, after a disaster, some level of federal assistance may be available to rebuild damaged bridges or roads, demolish and dispose of damaged buildings, rebuild schools or hospitals, or rebuild damaged levees. Such projects often have at least *some* impact on the environment, and hence may be required to comply with any of a number of local, tribal, state, or federal environmental laws—including requirements of the National Environmental Policy Act of 1969 (NEPA, 42 U.S.C. § 4321 et seq.).

Among other provisions, NEPA requires federal agencies to assess the potential environmental impacts of a proposed action before proceeding. Exempted from NEPA's requirements are emergency response actions under provisions of the Robert T. Stafford Disaster Relief and Emergency Assistance Act (the Stafford Act, 42 U.S.C. §§ 5121-5206). These exempted activities include providing essential relief to victims and implementing protective measures necessary to reduce immediate threats to life, property, and public health and safety. NEPA's environmental review requirements may, however, be applicable to long-term recovery projects, such as the modification, mitigation, or expansion of existing structures or the relocation of certain structures located in a floodplain.

Most agencies have implemented NEPA as an "umbrella" statute. As such, it forms a framework for the coordination or demonstration of compliance with any study, review, or consultation required by any other environmental law. The use of NEPA in this capacity has led to some confusion. The need to comply with a separate environmental law such as the Clean Water Act (CWA), Endangered Species Act (ESA), or National Historic Preservation Act (NHPA) may be *identified* within the framework of the NEPA process, but NEPA itself is not the source of the obligation. If, theoretically, the requirement to comply with NEPA were removed, compliance with each applicable law would still be required.

In the past, there has been congressional interest in the NEPA implementation process for disaster-related projects. This interest has been driven, in part, by federal grant applicants who have been confused about both their role in the NEPA process and what the law requires.

To address these issues, this report discusses the NEPA process as it applies to projects for which federal funding to recover from or prepare for a disaster may be requested by local, tribal, or state grant applicants. Specifically, the report provides an overview of the NEPA process as it applies to such projects, identifies the types of projects (categorized by the federal funding source) likely to require environmental review, and delineates both the types of projects for which no or minimal environmental review is required (those for which statutory or categorical exclusions apply) and those that likely require more in-depth review.

This report focuses on the NEPA process as it applies to projects that require grant applicants (i.e., state or local agencies) to provide certain information for their grant requests to be considered eligible for potential approval. It does not address the NEPA process as it applies to disaster-related projects that would likely involve collection of the necessary environmental review documentation by the federal agency responsible for the project (e.g., water resources projects undertaken by the Army Corps of Engineers). Two agencies that provide a significant proportion of applicant-requested funding for disaster-related projects are the Department of Homeland Security's Federal Emergency Management Agency (FEMA) and the Department of Housing and Urban Development (HUD)—particularly under its Community Development Block Grant (CDBG) program.[1] Therefore, this report primarily discusses the NEPA process as it applies to recovery and rebuilding projects funded under those agency programs.

Overview of the NEPA Process

NEPA is a procedural statute with twin aims that require agencies to consider the environmental impacts of their proposed actions and inform the public that environmental concerns have been accounted for in the decision-making process. The NEPA process involves the steps an agency must take to demonstrate that it has met these aims.

Environmental Review

NEPA requires all federal agencies to consider the environmental impacts of proposed federal actions before proceeding. Regulations that specify how agencies must implement NEPA's requirements were promulgated by the Council on Environmental Quality (CEQ).[2] CEQ regulations direct federal agencies to adopt and enforce their own regulations and procedures implementing NEPA's environmental review requirements in a manner specific to typical classes of actions undertaken by each agency. Two agencies that provide a significant proportion of funding for disaster-related projects are FEMA and HUD. Their NEPA regulations can be found at 44 C.F.R. § 10 (FEMA) and 24 C.F.R. 58 (HUD).[3] NEPA regulations specify environmental review requirements that must be met to demonstrate that potential environmental impacts have, in fact, been considered.

Generally, the term "environmental review" refers to a requirement to show evidence of formal consideration, evaluation, or analysis of the impacts of a proposed federal action. Most often, the use of the term is in reference to the process of complying with NEPA requirements. However, depending upon the project under consideration, an environmental review may refer to the process of identifying any environmental compliance requirements or exemptions, as applicable to a certain project.

As it has been interpreted, NEPA is a procedural statute that does not require agencies to elevate environmental concerns above others. Instead, NEPA requires only that an agency assess the potential environmental consequences of an action and its alternatives before proceeding. If adverse environmental effects of a proposed action are adequately identified and evaluated, the agency is not constrained by NEPA from deciding whether other benefits outweigh the environmental costs and moving forward with the action.

NEPA as an Umbrella Statute

Any given disaster-related project may require compliance with a wide variety of legal requirements enforceable by one or more agencies. For example, the environmental impacts of a given project may trigger compliance with elements of the Clean Air Act, Endangered Species Act of 1973, National Historic Preservation Act, or Clean Water Act.

Most individual agency NEPA procedures suggest that, for a given project, compliance with *all* applicable environmental laws, executive orders, and other legal requirements should be documented within the appropriate NEPA documentation. This concept is referred to as the "NEPA umbrella." As such, as previously noted, NEPA forms a framework for the coordination and demonstration of compliance with any study, review, or consultation required by other environmental laws. For example, the need to comply with another environmental law, such as the Clean Water Act, may be *identified* within the framework of the NEPA process, but NEPA itself is not the source of the obligation. If, theoretically, the requirement to comply with NEPA were removed, compliance with each applicable law would still be required.

An example of the use of NEPA as an umbrella statute can be seen in FEMA's environmental review process. Because of the types of projects the agency is likely to fund, and the high probability that historic properties may be affected in many projects, FEMA's NEPA process is referred to as an Environmental and Historic Preservation (E&HP) Review. In its description of the E&HP process, FEMA states:

> Federal environmental and historic preservation laws and Executive Orders provide the basis and direction for the implementation of federal environmental and historic preservation review requirements for FEMA-funded projects. These laws and executive orders are aimed at protecting our nation's water, air, coastal, wildlife, land, agricultural, historical and cultural resources, as well as minimizing potential adverse effects to children, low-income and minority populations. Failure to comply with these laws could result in project delays and denial of funding. FEMA through the Environmental and Historic Preservation Program engages in a review process to ensure that FEMA funded activities comply with these laws.[4]

Another example of the use of NEPA as an umbrella statute can be seen in the HUD Office of Community Planning and Development environmental review requirements applicable to CDBGs. Before a CDBG applicant can commit or expend funds for a given project, an environmental review of the project must be conducted. The environmental review record must, among other requirements, document compliance with applicable statutes and authorities. To meet this requirement, HUD regulations require the grantee to certify that it has considered compliance criteria applicable to historic preservation, floodplain management and wetland protection, coastal zone management, sole-source aquifers, endangered species, wild and scenic rivers, air quality, farmland protection, HUD environmental standards, and environmental justice.[5] That does not mean that all of these compliance factors will apply to a given project. The environmental review process is intended simply to identify the compliance requirements that *do* apply and ensure that the applicant will be compliant, as appropriate.

After Hurricane Katrina, some stakeholders cited NEPA as a significant challenge to state efforts to disperse CDBG funds.[6] The requirement to evaluate the various compliance criteria listed above was cited specifically as the problem. However, as stated previously, NEPA is not the source of these compliance requirements. The NEPA process simply forms the framework within which compliance with any applicable environmental law is *identified.* Still, this situation illustrates the difficultly some stakeholders have in distinguishing between what is required under NEPA and what may be required under other relevant environmental laws. It illustrates the challenges that applicants face when trying to comply with the range of requirements applicable to projects for which funding is sought.

Also, the comprehensive reviews, documentation, and analysis sometimes required by agencies such as the Army Corps of Engineers, the U.S. Fish and Wildlife Service, the Coast Guard, and the Environmental Protection Agency (EPA), as well as various state regulatory and review agencies (such as the office of a State Historic Preservation Officer), may add to the perception that project delays are related to the NEPA process. What may be perceived by the applicant as a NEPA-related delay may actually stem from an agency's need to complete a permit process or analyses required under separate statutory authority (e.g., the Clean Water Act or Endangered Species Act), over which the agency preparing the NEPA documentation has no authority.

NEPA Issues Relevant to Disaster-Related Projects

When a community is devastated by a disaster, it may be overwhelmed by the number of projects that need to be undertaken. Generally, when a local, tribal, or state agency participates in the NEPA process, it involves a project that has been planned or, at least to some degree, anticipated. In the wake of a disaster, however, when entire neighborhoods, towns, or regions may be substantially damaged and in need of repair or reconstruction, agencies can quickly become overburdened by the task of navigating applicable compliance requirements. Coupled with the potential difficulty of determining the various federal funding sources available to recover from or mitigate damages, local agencies responding to a disaster may become confused about their environmental compliance obligations. If those obligations are not met, funding will be slowed.

From the federal agency perspective, the commitment to assist in rebuilding structures and facilities and restoring land must be done in a way that will result in greater protection from future disasters. That is, federal agencies do not want to spend money that may have to be spent again when another disaster strikes. Elements of the NEPA process, such as the requirement to demonstrate flood-plain management considerations, help federal agencies meet this goal.

To understand what is required of applicants, it is helpful to understand the types of disaster-related projects associated with various funding sources and the levels of environmental review that may be required for proposed projects.

Disaster-Related Projects Potentially Subject to NEPA

NEPA's environmental review requirements apply to any project potentially subject to federal control or responsibility. Such actions include projects and programs entirely or partly funded, assisted, conducted, regulated, or approved by federal agencies.[7] With regard to disaster recovery, rebuilding, or mitigation projects, NEPA applies if federal funds are used for a project or if federal-agency permitting or other approvals are required.[8] Table 1 summarizes several potential funding sources for disaster-related projects, the types of projects potentially qualified for those funds, and the agencies authorized to provide funding.

Table 1. Projects and Funding Sources

Funding source	Types of projects[a]	Funding agency
Hazard Mitigation Grant Program (HMGP)	Projects that will reduce or eliminate the losses from future disasters, such as the acquisition of real property from willing sellers and demolition or relocation of buildings to convert the property to open space use; retrofitting of structures and facilities to minimize damages from high winds, earthquake, flood, wildfire, or other natural hazards; elevation of flood-prone structures; development and initial implementation of vegetative management programs; minor flood control projects; and localized flood control projects designed specifically to protect critical facilities.	FEMA

Funding source	Types of projects[a]	Funding agency
Public Assistance (PA) Grant Program	Projects intended to help a community respond to and recover from major disasters or emergencies declared by the President. Such projects include: debris removal; emergency protective measures; and the repair, replacement, or restoration of disaster-damaged, publicly owned facilities (such as roads, bridges, water control facilities, utilities, and critical buildings and equipment) and the facilities of certain private nonprofit organizations. Such projects are likely exempt from NEPA.	FEMA
Flood Mitigation Assistance (FMA) Program	Measures to reduce flood losses, such as elevation, acquisition, or relocation of National Flood Insurance Program (NFIP)-insured structures.	FEMA
Repetitive Flood Claims (RFC) grant program	Projects intended to reduce or eliminate claims under the National Flood Insurance Program (NFIP), such as the acquisition of properties, and either demolition or relocation of flood-prone structures, where the property is deed restricted for open space uses in perpetuity; building elevations; dry flood-proofing of non-residential structures; and minor localized flood control projects (funding limited to $1 million per project).	FEMA
Community Development Block Grant (CDBG) program	Short-term disaster relief, such as debris removal or the emergency restoration of essential services, such as water, sewer, electrical, and telecommunications; mitigation activities intended to lessen the impact of a future disaster, such as the construction of levees to protect against flooding, buildings designed to withstand earthquakes, or the buy-out of properties prone to a recurrence of disaster events; and long-term recovery activities, such as infrastructure improvements.	HUD

Source: Table prepared by CRS based on an evaluation of federal funding sources available for disaster response, recovery, and mitigation activities.

[a] This list is not intended to be exhaustive. In particular, it identifies categories of projects that would be initiated by local, state, or tribal agency grant applicants (as opposed to federal projects that may be initiated a federal agency, such as a flood-control project developed by the Army Corps of Engineers).

Agency and Applicant Roles

Generally, there are three entities that play a significant role in the NEPA process for disaster-related projects: the lead agency, which is responsible for preparing the NEPA documentation;[9] cooperating agencies, which may be any local, tribal, state, or federal agencies that have jurisdiction by law or special expertise regarding any environmental impact involved in a proposal;[10] and the project applicant (which may also be referred to as a responsible entity or grantee under different agency requirements), such as local, tribal, or state entities requesting federal funds.

For disaster-related projects, the lead agency coordinates environmental reviews for projects funded under that agency's programs. For example, FEMA would be the lead agency for debris removal operations involving Public Assistance (PA) Grant funds. In a disaster-stricken area, it is possible that multiple funding sources may be available for a single project. If more than one federal agency proposes or is involved in the same action, the agencies must determine which agency will serve as the lead and which will serve as cooperating agencies. If there is disagreement among agencies involved, factors including the magnitude of each agency's involvement and project approval/disapproval authority determine the designation of the lead agency.

A cooperating agency is one that has jurisdiction by law or special expertise regarding any environmental impact involved in a proposal.[11] It may be local, tribal, state, or federal agencies from which the lead agency will need technical assistance. For example, depending on the impacts associated with a given project, the following agencies may serve as cooperating agencies to develop appropriate NEPA documentation:

- The Environmental Protection Agency—to address air quality, water quality, and hazardous waste standards.
- Advisory Council on Historic Preservation and/or state or tribal historic preservation officer—to ensure that provisions of the National Historic Preservation Act are addressed.
- The Army Corps of Engineers—to ensure that any U.S. water or wetland impacts are identified, avoided or minimized, and mitigation resolved, among other potential requirements.
- The U.S. Fish and Wildlife Service—to evaluate projects to ensure that threatened and endangered species are identified, any impacts avoided or minimized, and mitigation resolved.

- The U.S. Department of Agriculture—to ensure that impacts to farmlands are considered.

Although the lead agency (such as FEMA or HUD) may be ultimately responsible for ensuring that NEPA documentation is complete, the project applicant plays a significant role in the NEPA process associated with disaster-related projects. The project applicant, such as a state or local agency, will likely be required to develop substantive portions of the environmental document, while the lead agency is responsible for its scope and overall content. For example, project applicants are required to provide information to support FEMA's Environmental and Historic Preservation compliance process. Funds will not be awarded, and the applicant may not initiate the project, until FEMA has completed its environmental review. Therefore, it is in the applicant's interest to gather and present all information necessary to assist the funding agency with meeting its environmental review requirements.

Categories of Action

Determining *whether* NEPA applies is generally not a complicated process. Determining what level of review is required for a project (i.e., whether its environmental impacts are significant) may not be as clear, and must be determined on a case-by-case basis. When determining NEPA's applicability, disaster-related projects will fall into one of the following categories:

- Statutory exemptions—applicable to certain FEMA-funded projects authorized under the Stafford Act.
- Categorical exclusions—applicable to actions that normally do not individually or cumulatively have a significant effect on the human environment and that the agency has determined from past experience have no significant impact.
- Environmental Impact Statements (EISs)—when the action will potentially have a significant environmental impact.
- Environmental Assessments (EAs)—when the significance of environmental impacts is uncertain and must be determined.

The requirement to produce an EIS is probably the most familiar element of NEPA compliance. However, actions requiring an EIS account for a small percentage of all federal actions proposed in a given year—and generally none of those associated with disaster response and recovery. For example, with regard to FEMA-funded projects, statutorily exempt and categorically excluded projects account for approximately 99% of actions undertaken annually.[12] Further, from 2000 through 2008, FEMA filed *no* EISs.[13]

When a project that involves potentially significant effects is implemented under emergency conditions, alternative compliance arrangements may be considered when determining how an environmental review may be carried out. These alternative arrangements and the various categories of action potentially subject to NEPA are discussed below.

Statutory Exemptions

In responding to emergencies and major disasters, existing provisions of the Stafford Act statutorily exempt certain FEMA-funded activities from NEPA. Statutory exclusions generally apply to actions that are emergency in nature or are necessary for the preservation of life and property. They apply to most Public Assistance actions funded by FEMA, but do not apply to hazard mitigation, flood mitigation, unmet needs projects, or FEMA grant programs. Specifically, response actions excluded from NEPA by the Stafford Act (at 42 U.S.C. § 5159) include the following:

- General federal assistance—such as the utilization of personnel, equipment, supplies, technical and advisory services in support of disaster assistance. (See actions specified under 42 U.S.C. § 5170a.)
- Essential federal assistance—including actions to meet immediate threats to life and property resulting from a major disaster, such as: the use of federal resources (e.g., equipment, supplies, and facilities), medicine, food, and other consumables; and work and services to save lives and protect property (e.g., debris removal, search and rescue, clearance of roads, demolition of unsafe structures, warning of further risks and hazards). (See actions specified under 42 U.S.C § 5170b.)
- Repair, restoration, and replacement of damaged buildings—generally, this means restoring the facility to the same function, capacity, and footprint. (See actions specified under 42 U.S.C. § 5172.)
- Debris removal—exempt debris removal activities include clearance of debris and wreckage resulting from a major disaster from publicly

and privately owned lands and waters after a major disaster (long-term debris removal activities, such as decisions on landfill locations, may not be exempt from NEPA). (See actions specified under 42 U.S.C. § 5173.)

- Federal emergency assistance—such as the utilization of personnel, equipment, supplies, technical and advisory services in support of disaster assistance; and assistance in support of medicine, food, and other consumable supplies and emergency assistance. (See actions specified under 42 U.S.C. § 5192.)

It is important to understand that, as with actions categorically excluded (discussed below), an action statutorily excluded from NEPA is not exempt from the requirements of *other* environmental statutes. FEMA would still be responsible for complying with all other applicable local, state, tribal, and federal laws and regulations relating to health, safety, and the environment.[14] This could encompass federal environmental statutes including, among others, the Clean Air Act, Clean Water Act, Endangered Species Act, National Historic Preservation Act, Resource Conservation and Recovery Act, Coastal Zone Management Act, and the Coastal Barrier Resources Act.

Also, these exemptions are specific to designated FEMA-funded activities authorized under the Stafford Act. Similar statutory exemptions do not exist for HUD-funded programs such as CDBG disaster assistance. In fact, under the CDBG funding program, states are allowed to seek waivers of certain program requirements, *except* those related to environmental review (and fair housing, nondiscrimination, and labor standards).

Categorical Exclusions

If a project is of a type that falls within an established category of activities the agency has previously determined to have no significant environmental impacts, it is categorically excluded from the requirement to prepare an EA or EIS. Sometimes such actions are referred to as being categorically excluded or exempt from NEPA. However, NEPA *does* apply to such actions; they are excluded only from the requirement to prepare an EA or EIS.

Individual agencies are required to specifically list, in their respective NEPA regulations, those projects likely to be considered categorical exclusions.[15] For example, FEMA has identified, among others, the following actions as generally classifiable as categorical exclusions: upgrades to codes and standards, removal of structures after addressing historic preservation needs, and minor improvements or minor hazard mitigation measures at existing facilities, such as placing riprap at a culvert outlet to control erosion.[16] An example of a HUD-identified categorical exclusion is the "acquisition, repair, improvement, reconstruction, or rehabilitation of public facilities and improvements (other than buildings) when the facilities and improvements are in place and will be retained in the same use without change in size or capacity of more than 20 percent (e.g., replacement of water or sewer lines, reconstruction of curbs and sidewalks, repaving of streets)."[17] Note that this categorical exclusion is similar to the Stafford Act's statutory exemption for projects that would repair, restore, or replace damaged buildings. Most agencies have identified similar activities in their list of categorical exclusions.

Whether or what types of documentation may be required to demonstrate that a project is categorically excluded will depend on whether the project involves extraordinary circumstances that may cause a normally excluded action to have a significant environmental effect.[18] Also, the fact that a project does not have a significant impact as defined under NEPA does not mean that it will not trigger statutory requirements of other environmental laws. For example, if historical sites, endangered species habitat, wetlands, or property in minority neighborhoods, to name a few, would be affected by a proposed federal action, compliance with related environmental laws or requirements, in addition to NEPA, may be required.

Even though categorically excluded projects do not have significant environmental impacts, an agency may require a certain level of documentation to prove that the CE determination is appropriate. For example, FEMA's NEPA regulations identify three levels of categorical exclusions and the types of documentation necessary for each.[19]

If there are unresolved extraordinary circumstances that may have a significant adverse environmental impact, such as the potential to affect protected natural or cultural resources, the proposed action cannot be categorically excluded, and an EA is required.

Projects Requiring an EA or EIS

If a project is not statutorily exempt from NEPA or does not fit the criteria applicable to a categorical exclusion, it must be determined whether the environmental impacts of such a project will be significant, and hence require the preparation of an EIS. An EIS may be required for projects intended to facilitate long-term recovery of an affected region. Examples may include disaster-related flood-control or hurricane protection projects (e.g., new wetlands restoration projects or levee repair projects); construction of roads, bridges, storm water management projects, tornado shelters, temporary housing, fuel modification projects, and public facilities (e.g., schools, libraries, utilities); debris storage, staging, and removal; and building acquisition, relocation, and demolition.

If past disasters can serve as a gauge, a disaster-related project likely will not require an EIS. However, it may require an EA. An EA is carried out in order to clarify issues and determine the extent of an action's environmental effects. CEQ regulations define an EA as a concise public document that serves to briefly provide sufficient evidence and analysis for determining whether to prepare an EIS or a finding of no significant impact (FONSI); aid agency compliance with NEPA when no EIS is required; and facilitate preparation of an EIS when one is necessary.[20]

The CEQ regulations require no standard format for EAs, but do require agencies to include a brief discussion of the need for the proposal, alternatives, impacts of the proposal and alternatives, and a list of agencies or individuals consulted.[21] Individual agency regulations and/or guidance may include more specific requirements. Some agencies suggest that the process for developing an EA should be similar to the process for developing an EIS. For example, the applicant should consult interested agencies to scope the project to determine the potential for social, economic, or environmental impacts; briefly discuss the project's purpose and need; identify project alternatives and measures to mitigate adverse impacts; and identify any other environmental review requirements applicable to the project (e.g., permitting requirements under the Clean Water Act). Public participation in the EA process is left largely to the discretion of the lead agency.

If at any time during preparation of the EA it is determined that a project's impacts are significant, EIS preparation should begin. If it is ultimately determined that impacts are not significant, the lead agency must prepare a FONSI. The FONSI serves as the agency's administrative record in support of its decision regarding a project's impact. The FONSI also must be available to the affected public.[22]

Alternative Compliance Arrangements

In addition to categorical and statutory exclusions to NEPA, both CEQ and individual agencies have specified "Alternative Arrangements" for complying with NEPA's requirements in the event of an emergency.[23] Alternative arrangements are available where emergency circumstances make it necessary to take an action with significant environmental impacts (i.e., a project that would otherwise require an EIS) without observing the provisions of the applicable NEPA regulations. In such circumstances, the federal agency taking the action should consult with CEQ about what those arrangements may be and the time frame within which they must be completed. Exactly what those arrangements involve and how they would be implemented would vary according to the nature of the disaster. Generally, their intent is to expedite the NEPA process when an EIS would otherwise be required.

These Alternative Arrangements do not waive the requirement to comply with NEPA regulations, but establish an alternative means of compliance. Agencies and CEQ are to limit such arrangements to actions necessary to control the immediate impacts of the emergency.

On September 8, 2005, CEQ released a memorandum that provides guidance on emergency alternative arrangements under NEPA. That guidance was specific to Hurricane Katrina,[24] but could be applicable to any disaster response activities undertaken by federal agencies. For example, CEQ lists activities that could be considered for analyses in accordance with alternative arrangement provisions, such as the disposal of unsorted disaster debris (waste that includes both hazardous and nonhazardous constituents) at a specific site or the permanent replacement of certain major facilities when the agency expects that significant environmental affects will occur.

On March 23, 2006, FEMA published a notice specifying those arrangements for agency grants to repair or reconstruct critical infrastructure in the New Orleans Metropolitan Area.[25] "Critical infrastructure" included

- hospitals and health care facilities,
- utilities and wastewater treatment plants,
- permanent police and fire stations,
- government and court administration buildings,
- detention centers (jails), and
- permanent schools.

Although alternative arrangements are an option after a disaster, they are rarely used. Since 1980, the only disaster-related projects to use alternative arrangements involved certain rebuilding efforts in the New Orleans area after Hurricanes Katrina and Rita. In particular, alternative arrangements were established for the Corps' reconstruction of levies damaged after the storm and FEMA grant-assisted projects to replace certain critical infrastructures (referenced above). Specifically, after the hurricanes, FEMA anticipated that grant applications from Louisiana "will more strongly reflect future demands than returning to pre-disaster conditions. Proposed projects will not necessarily be the same size, nature or location; will use current building codes, and construction methods; and take advantage of current community and urban planning principles, and hazard mitigation opportunities."[26] Under such conditions, NEPA *would* apply. To coordinate their recovery efforts more quickly, among other factors, FEMA requested Alternative Arrangements to meet its NEPA requirements.

CONCLUSION

When a community is devastated by a disaster, all at once it may be faced with the need to rebuild roads, bridges, private homes, and public buildings. Such projects, when undertaken under normal circumstances, may not qualify for federal funding, but when they do, as in the case after a disaster, the federal government must be assured that certain criteria are met before those funds are made available. Among other factors, the federal government needs confirmation that those funds will not be used in a way that increases the likelihood that their investment will be lost if and when another disaster strikes the same area. Further, before a federal agency will provide funds for disaster recovery or rebuilding, it must gain assurance that the project complies with applicable regulations, laws, and executive orders. The NEPA process is a vehicle by which that assurance can be obtained.

End Notes

[1] For more information about the various grant programs available to assist local governments with disaster recovery projects, see FEMA's "Grants and Assistance Programs for Governments" at http://www.fema.gov/government/grant/ government.shtm and "CDBG Disaster Recovery Assistance" at http://www.hud.gov/offices/cpd/communitydevelopment/ programs/drsi/index.cfm.

[2] 40 C.F.R. §§ 1500-1508.

[3] Additional environmental review requirements, potentially applicable to CDBG projects, are specified under 24 C.F.R. § 51(Environmental Criteria and Standards and 24 C.F.R. § 55 Floodplain Management).

[4] Available on FEMA's "Environmental & Historic Preservation Review" site at http://www.fema.gov/plan/ehp/ ehpreview/index.shtm.

[5] 24 C.F.R. § 58.5.

[6] Hearing before Senate Committee on Banking, Housing and Urban Affairs, "Two Years After the Storm: Housing Needs in the Gulf Coast," Statement of Edgar A.G. Bright, III, CMB President, Standard Mortgage Corporation Member of the Residential Board of Governors of the Mortgage Bankers Association, September 25, 2007. Available at http://banking.senate.gov/public/index.cfm?FuseAction=Hearings.List&Month=0&Year=2007.

[7] 40 C.F.R. § 1508.18(a). Further, the term "federal agency" is defined as all agencies of the federal government, but does not mean the Congress, the Judiciary, or the President (40 C.F.R. § 1508.12).

[8] If a project is undertaken purely by using city or state funds and it requires no federal authorization or permit, NEPA's environmental review requirements do not apply. However, depending on the nature of the project, compliance with other environmental laws may be required. Also, some cities and states have their own NEPA-like requirements.

[9] 40 C.F.R. § 1508.16.

[10] 40 C.F.R. § 1508.5.

[11] 40 C.F.R. § 1508.5.

[12] See FEMA's *Coordinating Environmental and Historic Preservation Compliance* training materials, "Module 2, Lesson 2: FEMA Exclusions Under NEPA," p. 1, January 2004, available at http://www.training.fema.gov/EMIWeb/ IS/IS253lst.asp.

[13] The total number of EISs filed annually, by agency, is posted by CEQ at http://ceq.hss.doe.gov/nepa/nepanet.htm.

[14] Local, state, and federal environmental laws and regulations may still provide some exemption or regulatory allowance in the event of an emergency.

[15] 40 C.F.R. § 1507.3.

[16] 44 C.F.R. § 10.8(d)(3).

[17] 24 C.F.R. § 58.38.

[18] 40 C.F.R. § 1508.4.

[19] See "FEMA Categorical Exclusions (CATEX)" at http://www.fema.gov/plan/ehp/ regionviii/catex.shtm.

[20] 40 C.F.R. § 1508.9(a).

[21] 40 C.F.R. § 1508.9(b).

[22] 40 C.F.R. § 1501.4(e)(1).

[23] 40 CFR § 1506.11. The Corps has regulatory provisions similar to CEQ's that address emergency actions. FEMA's regulations reference statutory exemptions provided under the Stafford Act.

[24] See "Memorandum for Federal NEPA Contacts: Emergency Actions and NEPA," including Attachments 1 and 2, at http://www.nepa.gov/nepa/regs/guidance.html.

[25] 71 *Federal Register* 14712-14716; see also FEMA's webpage "Greater New Orleans Infrastructure Projects," at http://www.fema.gov/plan/ehp/noma/about1.shtm.

[26] 71 *Federal Register* 14715.

In: The National Environmental Policy Act ISBN: 978-1-61324-876-8
Editors: M. Drubich and A. Moser

Chapter 4

THE 2010 OIL SPILL: MMS/BOEMRE AND NEPA*

Kristina Alexander

SUMMARY

On April 20, 2010, an exploratory oil well in the Gulf of Mexico exploded, killing 11 people and causing the worst oil spill in American history. The oil well was on a tract leased by BP, having obtained a lease and the relevant permits from the federal government. Under relevant federal law, federal actions that may have adverse environmental effects are required to be reviewed for potential environmental harm under the National Environmental Policy Act (NEPA). This report will review those environmental procedures. While there are additional environmental obligations imposed on Outer Continental Shelf (OCS) drilling by other acts, this report will not review those requirements.

Multiple environmental reviews were conducted by the Minerals Management Service (MMS) at each stage of OCS development. (MMS was reorganized in May 2010, and the relevant office for the environmental reviews is the Bureau of Ocean Energy Management, Regulation, and Enforcement (BOEMRE). This report will refer to MMS for actions before the reorganization and BOEMRE going forward.) For the particular well in question, MMS addressed the environmental impacts on four occasions, including two full environmental impact

* This is an edited, reformatted and augmented version of a Congressional Research Service Report R41265, dated March 2, 2011.

statements, an environmental assessment, and a categorical exclusion. The fact that MMS categorically excluded the exploration plan from a NEPA analysis is controversial, since that is the first step in which drilling would be conducted. It appears MMS followed its internal procedure for NEPA reviews in the western and central areas of the Gulf of Mexico by employing a categorical exclusion for an exploration plan. However, that procedure has never been reviewed by a court to see if it is consistent with the law or whether an exception to the categorical exclusion may apply in this case. Had this project occurred in a different geographical area, including the eastern area of the Gulf of Mexico, it likely would have undergone a higher level of environmental scrutiny. Following a White House report suggesting that BOEMRE review its NEPA exclusions, BOEMRE announced it would not apply the categorical exclusion used in the Gulf until its review was complete.

Congress has addressed the issue of MMS/BOEMRE categorical exclusions in proposed legislation. In the 112th Congress, bills have been introduced that would require either a full environmental impact statement (H.R. 52 (Connolly)), or at least an environmental assessment (H.R. 501, §§ 208, 215 (Markey)) for exploration, development, and production plans. Both would change the 30-day deadline for approval of exploration plans. Another proposed NEPA change would prevent the act from applying to testing oil spill prevention, response, or mitigation technology in Arctic waters and bar judicial review (S. 203 (Begich)). In the 111th Congress, the Consolidated Land, Energy, and Aquatic Resources Act of 2010 (CLEAR) (H.R. 3534) would have required certain levels of environmental reviews for exploration, development, and production plans, and it would have changed the 30-day statutory deadline for approving exploration plans to 90 days.

Background and Context

On April 20, 2010, an exploratory oil well in the Gulf of Mexico exploded, killing 11 people and causing the worst spill in United States history at approximately 4.1 million barrels (bbl).[1] It was capped in mid-July. The oil well was on a tract leased by BP, having obtained an oil and gas lease and the relevant permits from the federal government. The spill was at the Mississippi Canyon Block 252, which is one block obtained by BP under Lease Sale 206. Under applicable federal law, federal actions that may have adverse environmental effects are required to be reviewed for potential environmental harm under the National Environmental Policy Act (NEPA). Additional environmental obligations are imposed by the Outer Continental Shelf Lands

Act (OCSLA),[2] Endangered Species Act,[3] Marine Mammal Protection Act,[4] and the Coastal Zone Management Act[5] but will not be examined within this report. Rather, this report focuses on the environmental analyses conducted by the Minerals Management Service (MMS) and their compliance with NEPA. MMS was reorganized in May 2010,[6] and the relevant successor office is the Bureau of Ocean Energy Management, Regulation, and Enforcement (BOEMRE). This report will refer to MMS for events prior to the reorganization.

NEPA

NEPA states that for all "major Federal actions significantly affecting the quality of the human environment" federal agencies shall prepare a "detailed statement" of "the environmental impact of a proposed action, any adverse environmental effects which cannot be avoided should the proposal be implemented, [and] alternatives to the proposed action."[7] Agencies must comply to "the fullest extent possible."

MMS/BOEMRE of the Department of the Interior is the federal agency charged with overseeing oil and gas exploration of the Outer Continental Shelf (OCS). MMS issued the lease to and approved the exploration plan of BP for the oil well that caused the spill. Under NEPA terminology, MMS/BOEMRE is the *lead agency* for any environmental reviews.[8]

NEPA includes three types of environmental reviews with different levels of scrutiny into the environmental effects of an action. The goal of the reviews is for an agency to be able to demonstrate that it has taken an appropriately *hard look* at the environmental consequences of its planned activity.[9] The environmental review is also supposed to involve the public.[10] The most comprehensive review is concluded by an environmental impact statement (EIS). It is required by NEPA for all agency actions that will significantly affect the environment. An EIS should demonstrate that the agency considered a reasonable range of alternatives that took a hard look at the environmental consequences of a proposed action. An EIS also requires extensive public involvement, including a public comment period for a draft EIS and consideration of those comments in the final document.

Where an action may not have significant effects, or when an agency is unsure of the degree to which an action may have significant impacts, a review called an environmental assessment (EA) is conducted. An EA also requires consideration of alternatives to the action and a review of the effects, but it is intended to be an abbreviated review. The public comment period is limited and could occur only upon completion of a final document.

The third type of review is for actions that agencies have pre-determined have no significant impact, typically because an agency routinely conducts the activity. This type of review is a categorical exclusion (CE, sometimes known as a Cat Ex). Technically, it means a formal review is not required because the agency has already determined that the environmental consequences of the action will not be significant.[11] Agencies prepare lists of CEs and may exclude a particular action if it is on that list. The decision to invoke a CE for a project must be documented in some way.[12] CEs are not allowed when there are extraordinary circumstances surrounding a proposed action, which generally include the presence of endangered species or significant resources of some type, such as archeological sites. This is consistent with the purpose of CEs—they do not apply when there could be a significant environmental impact.

Aside from the interpretive case law, the requirements for NEPA reviews can be found in two places: the regulations issued by the Council on Environmental Quality (CEQ);[13] and within agency-specific regulations or guidelines. Generally speaking, the CEQ regulations provide the foundation for NEPA compliance, while the agency guidelines provide for more specific application to the circumstances of a particular agency. In the case of MMS/BOEMRE, NEPA procedures are found within the Department of the Interior agency-wide NEPA guidelines in the Departmental Manual (DM),[14] as well as in MMS guidelines within the DM.[15]

MMS/BOEMRE NEPA Reviews

MMS/BOEMRE's summary of its OCS program indicates there are four different stages at which it performs an environmental review for OCS leasing.[16] The summary indicates what type of environmental review is typically done at each stage. Those stages, and the typical environmental reviews, are

- develop a 5-year program—EIS;
- plan for a specific lease sale—EIS;
- approve the exploration plan—EA; and
- approve development and production plan—EIS.

The four stages are based on the OCSLA.[17]

Records indicate that MMS conducted four separate environmental reviews related to the oil well known as Mississippi Canyon Block 252, including two EISs, one EA, and a CE. These four documents relate to the first three stages described above; two reviews were conducted at the lease sale stage, and the development and production stage had not been reached. Mississippi Canyon Block 252 was at the exploration stage. Briefly, those four documents are as follows:

- April 2007: MMS issued an EIS for the OCS Five Year Leasing Program, the 2007-2012 Outer Continental Shelf (OCS) Oil and Gas Leasing Program. This document considered environmental impacts related to off shore drilling, not just within the Gulf of Mexico, but for the two other areas in which off shore drilling is conducted: around Alaska and along the Atlantic coast.
- April 2007: MMS issued an EIS for the lease sale in the western and central portion of the Gulf of Mexico.
- October 2007: MMS issued an EA for Lease Sale 206, which included Mississippi Canyon Block 252, finding no new significant impact. This document was intended by MMS to supplement the other environmental reviews, which is why instead of the accepted NEPA terminology of "finding of no significant impact" (FONSI) it referred to no "new" significant impacts.
- April 2009: MMS issued a CE for the exploration plan that authorized BP to begin exploratory drilling on its site.

Additionally, two other environmental reviews were conducted that relate to drilling in the Gulf. Both were programmatic environmental assessments, meaning they were designed to consider broad impacts from a type of federal action and not just the impacts from one action. Programmatic environmental reviews are viewed as an efficient way to consider impacts without creating redundant reviews.[18] In July 2004, MMS announced completion of a programmatic EA that evaluated potential impacts of geological and geophysical testing in the Gulf.[19] That testing included seismic testing. A different programmatic EA was announced in March 2005 for Structure-Removal Operations in the Gulf,[20] which typically involve the use of explosives.

The 2007-2012 Outer Continental Shelf Oil and Gas Leasing Program

In February 2006, and again in August 2006, MMS sought comments on its proposed 5-year lease plan for 2007-2012 that would expand OCS leasing areas. The proposed plan addressed 21 OCS areas, including two in the Gulf of Mexico.[21] An EIS for this plan was prepared, and the final 5- year Plan EIS was completed in April 2007.[22] The 2007-2012 OCS Oil and Gas Leasing Program became effective on July 1, 2007. However, a federal court of appeals determined that MMS had not complied with all of the environmental requirements, and the program was returned to MMS for revision.[23] A preliminary revised 5-year plan was announced in March 2010.[24]

The EIS for the 5-year Plan

The EIS for the 5-year Plan for 2007-2012[25] describes its scope as being regional and programmatic, and it also describes the assumptions made when evaluating the environmental effects of the proposed action. It divides the environmental effects based on the three geographical areas within the 5-year Plan: Gulf of Mexico, Atlantic, and Alaska. Different assumptions are made for each area, and the environmental effects are premised on that. One assumption for the Gulf of Mexico is the scale of oil spills that could be expected to occur.[26] The estimates were made for potential spills during production and transportation, and do not appear to consider oil spills at the exploration stage. A large oil spill, according to the 5-year Plan EIS, would be one of 1,000 barrels (bbl) or greater. Based on historical modeling, MMS calculated a probable large oil spill from a platform to be 1,500 bbl and 4,600 bbl from a pipeline.[27] The spill from the current Mississippi Canyon exploration well is categorized as a platform-related spill, even though the leak occurred far below the surface of the water. The 5-year Plan EIS estimates that the Gulf of Mexico wells would produce 4 to 8 billion bbl of oil and have nine large oil spills (four from pipelines; four from platforms; and one from a tanker).[28] These numbers are for deepwater and near-shore drilling combined. Approximately 75% of the leasing activity is planned for deepwater.

The programmatic environmental effects for the preferred alternative for the Gulf Coast included impacts on the following: air and water quality; mammals and birds; fish, fisheries, and fish habitat; sea turtles; coastal and seafloor habitats; areas of special concern; socioeconomic impacts;

archeological resources; tourism and recreation; and land use.[29] Some protected resources in the area include the following endangered species:

- Northern Right Whale,
- Blue Whale,
- Fin Whale,
- Sei Whale,
- Humpback Whale,
- Sperm Whale,
- West Indian Manatee,
- Leatherback turtle,
- Green turtle,
- Hawksbill turtle,
- Kemp's Ridley turtle,
- Loggerhead turtle,
- Gulf Sturgeon,
- Whooping crane,
- Piping plover,
- Alabama beach mouse,
- Choctawhatchee beach mouse,
- St. Andrew beach mouse, and
- Perdido Key beach mouse.

The baseline data described above largely influence the discussion of environmental impacts within this EIS. The data contemplate four platform spills across the entire 40-year lease term (and not just the five years of the 5-year Plan) that would produce spills of approximately 1,500 bbl each, and only three of those spills would be in deepwater. Accordingly, the analysis of the effects from the spills is limited to considering how a 1,500 bbl spill could affect the environment. There is no extrapolation or other hypothesis for what would happen if the spill were larger. Based on the assumptions of limited spill quantities, the conclusion for the environmental consequences tends to be in one of two camps: one, that the magnitude of harm would depend on the timing and quantity of the oil spill (see, for example, Coastal Waters, Marine Waters, Marine Mammals, Marine and Coastal Birds, Essential Fish Habitat, Sea Turtles); or two, that because a deepwater spill is calculated to be at 1,500 bbl, it is unlikely to reach areas where significant impacts could occur (see, for example, Marine and Coastal Birds, Terrestrial Mammals, Coastal Barrier

Beaches, Sea turtle nesting, Seafloor Habitat and Live Bottom and Pinnacle Areas).

Environmental groups argued that the agency failed to take a hard look at the environmental consequences of its action. Limiting the size of a potential spill could limit the size of impacts that are reviewed. However, the court that reviewed this matter, the D.C. Circuit Court of Appeals,[30] has determined that challenges to the sufficiency of that NEPA process for the 5-year plan could not be reviewed.[31] The court said that the 5-year Plan NEPA review had "not yet reached that 'critical stage' where an 'irreversible and irretrievable commitment of resources' has occurred that will adversely affect the environment."[32] The court suggested that the appropriate time for such claims would be at the leasing stage. Accordingly, this ruling bars any challenge to the 5-year Plan NEPA review.

This position is consistent with other courts that have reviewed NEPA challenges of multiple stage administrative programs. The Second Circuit has held that a federal action, in that case an administrative hearing involving Mobil Oil, is not subject to NEPA until that point where there might be irreversible and irretrievable commitments of resources.[33] Similarly, the D.C. Circuit held that a Forest Service procedure for identifying areas for oil and gas leases was not ripe because it could prove unnecessary if no leases were ever issued.[34] No irreversible or irretrievable commitment of resources had been made.

The EIS for 11 Gulf of Mexico Lease Sales

In April 2007, MMS also completed the EIS for the 11 lease sales for the Gulf of Mexico, the Multisale EIS.[35] For the lease sales, the MMS divided the Gulf of Mexico into three regions: western (the Texas coast); central (the coasts of Louisiana, Mississippi and Alabama); and eastern (the Florida coast). Eleven lease sales were planned for the western and central areas and none for the eastern area.[36] A number of leases could be issued within each of the 11 lease sales, typically over 300, based on the amount of resources an area is estimated to produce. For more details on the facts underlying the western and central lease sales, see Appendix A.

The Multisale EIS evaluated the environmental impacts from routine operations as well as accidents such as oil spills. It reviewed all of the different steps involved in extracting oil— seismic testing, exploratory drilling, development and production, transportation, and removing structures at the end of the lease.

Blowouts and other spills were evaluated, first, on the probability of such events occurring, and then analyzing the probable resulting environmental effects. Like the 5-year Plan EIS, the Multisale EIS used historical data as the basis for the size of an oil spill. According to MMS, the most likely size of an oil spill that was greater than 1,000 bbl would be 4,600 bbl.[37] However, MMS states that large oil spills (meaning spills greater than 1,000 bbl) are "low-probability events."[38]

MMS used modeling to see how a 4,600 bbl spill would dissipate when on the surface. (Although it studied both surface and subsurface spills, MMS noted that subsurface spills would behave the same as surface spills once they reached the top.) MMS found that natural weathering would dissipate 32% to 74% of the slick; between 30% and 32% would be lost to the atmosphere via evaporation; and about 2% to 42% would be lost into the water column via natural dispersion.[39] No modeling was done on a larger spill.

The Multisale EIS did consider the possibility of a larger offshore spill for the leases, although it did not analyze impacts from such a spill. It found that a spill larger than 10,000 bbl had greater than a 99% chance of occurring during the 40-year period.[40] A mean number of spills of that size was estimated between 11 and 13 for that time period.[41] MMS analyzed the environmental effects of a spill of 4,600 bbl, the "most likely size of a spill greater than 1,000 bbl."[42]

In general, the probability of an offshore oil spill greater than 1,000 bbl (but not larger than 4,600 bbl) reaching an environmentally sensitive resource was found to be small, ranging from less than 0.5% (for example, reaching Gulf communities, listed beach mice, or Gulf Sturgeon habitat), but greater for other resources (for example, the probability of such a spill reaching waters used by coastal sea turtles ranged from 6% to 35%, while the odds of a spill reaching sea turtle nesting and mating habitat were 4%).[43] Thus, an offshore spill larger than 1,000 bbl was not found to be a significant environmental impact.

The Multisale EIS includes a discussion of blowouts as a *Loss of Well Control* (LWC), of which a blowout is the most severe form. Most LWC events were found to last half a day.[44] Therefore, MMS concluded impacts to marine water quality were not estimated to be significant.[45] However, it is noted that the noise from a blowout explosion could injure marine mammals, depending on their proximity.[46]

Adequacy of the Environmental Review of the Multiple Lease Sales

It could be questioned whether the review within the Multisale EIS was comprehensive enough to satisfy a court that MMS fully complied with NEPA. The act requires “a detailed statement” on “any adverse environmental effects which cannot be avoided should the proposal be implemented.”[47] A court could consider the adequacy of not reviewing the environmental effects of spills over 10,000 bbl when MMS calculated a greater than 99% probability that such a spill would occur. On the other hand, the court could find that reviewing spills of 4,600 bbl met NEPA’s standards. There is no court decision reviewing this document.

NEPA requires an agency to consider the environmental consequences of its actions during the development of a proposal.[48] Courts apply the “rule of reason” to see if the document is sufficient. One court described that review as finding whether an EIS “contains a reasonably thorough discussion of the significant aspects of the probable environmental consequences.”[49] In general, courts have held that the goals of NEPA are satisfied when an EIS “sets forth sufficient information to enable the decision-maker to consider fully the environmental factors involved.”[50]

In the case of OCS leases, courts have accepted that there are staged analyses with different degrees of scrutiny matching the different OCSLA steps. In a suit challenging an EIS prepared for one lease sale in the Arctic,[51] the plaintiffs argued that the NEPA analysis was flawed because it significantly underestimated the amount of oil that might be spilled.[52] The Ninth Circuit held that they were “least troubled by what may seem to be incomplete or speculative data at the lease sale stage,” noting that prior to exploration “an oil spill risk analysis can never be more than speculative.”[53] The court’s comfort with the data was based on the fact that OCSLA had tiered analyses: “the amount and specificity of information necessary to meet NEPA requirements varies at each of OCSLA’s stages.”[54] It referred to the U.S. Supreme Court decision regarding the tiered environmental analysis of OCSLA, *Secretary of the Interior v. California*.[55] In that case, which did not consider NEPA, the Supreme Court said that a lease sale on its own “does not directly mandate further activity that would raise an oil spill problem,” and therefore, there was no need to perform a CZMA environmental analysis at that point.[56] The Supreme Court and the Ninth Circuit predicated their decisions on the fact that the desired environmental review could occur at a later stage in the OCSLA process.

Other courts have considered the argument that an environmental document at the lease sale stage should have considered the potential effects of a larger spill, ultimately rejecting that claim.[57] In a dispute in the Ninth Circuit, the NEPA document had discussed impacts from spills exceeding 1,000 barrels and those exceeding 10,000 barrels for OCS drilling in the Arctic. The plaintiff wanted consideration of spills of greater than 100,000 bbl, based on the Amoco Cadiz spill of 1.6 million barrels in 1978, and the 3.5 million bbl exploratory well blowout in the Gulf of Mexico in Mexican waters in 1979. The district court held that plaintiff had not made its argument well, failing to show how impacts from a 100,000 bbl spill would be different from those discussed for a 10,000 bbl spill. Additionally, the court noted that an EIS does not have to consider every possible impact: "The rule of reason applicable to environmental impact statements does not require an inquiry into every conceivable situation that may occur no matter how remote or speculative it may be."[58]

The Ninth Circuit agreed with the conclusion of the district court, but for different reasons. It found that the tiered environmental reviews under the OCSLA would mean that as drilling became more likely, the impacts of a 100,000 bbl spill may be considered. The lease sale stage was a preliminary stage, according to the court: "Further information about the probability and location of a 100,000 barrel spill will become available as lessees survey their tracts, or test them, or plan for production and development."[59]

While NEPA requires consideration of a range of alternatives, it does not require consideration of a worst case scenario, such as another 3.5 million bbl spill.[60] An earlier version of the CEQ regulations did require such an evaluation when an agency had "incomplete or unavailable information" regarding the effects of an action. In that case, an agency was required to "include a worst case analysis and an indication of the probability or improbability of its occurrence."[61] In general, courts have held that an EIS "is not required to discuss every conceivable consequence of an agency's actions. Under the rule of reason, the environmental impact statement is not required to consider alternatives or consequences that are only speculative or are too remote."[62] It would be up to a court to consider whether a blowout spill from Lease Sale 206 was too remote and speculative for MMS to consider in its environmental review.

Lease Sale 206 Environmental Assessment

The third environmental review in the OCS process is the environmental assessment (EA) prepared for Lease Sale 206 in the Gulf (Lease Sale 206

EA).[63] The oil spill occurred in Mississippi Canyon Block 252 of this lease sale. MMS prepared environmental assessments for other particular lease sales in the western and central Gulf, such as Lease Sale 207, although a Supplemental EIS was prepared for Lease Sale 224 in the eastern Gulf.

Typically, an EA leads to one of two conclusions: discovery of a significant impact and then preparation of an EIS, or a finding of no significant impact (FONSI). MMS terms its EAs differently. It issues an EA-FONNSI, meaning finding of no *new* significant impact, reflecting the tiered environmental analyses for the OCS process. Lease Sale 206 EA recognizes that it "tiers off the Multisale EIS and incorporates much of the material by reference."[64] According to MMS, new information related to the lease sale was discovered, but did not change the conclusions reached within the Multisale EIS.[65]

The EA appears to back off some of the environmental impacts considered in the EIS. It states that because activity was overestimated, the environmental impacts "may have been overstated."[66] While new information was considered within the EA, MMS found it supports the conclusions made in the Multisale EIS, and so no new conclusions are made within the EA.[67] The Lease Sale 206 EA refers to the Multisale EIS review of spills greater than 1,000 bbl.[68] No reference to larger spills was found, such as a 4,600 bbl spill. In fact, oil spills were discussed in more general terms throughout the document than in either of the EISs, which were incorporated by reference.

No legal challenge to the Lease Sale 206 EA has been found. However, challenges to other EAs for OCS drilling have been found for EAs issued for particular lease sales in the Arctic. Courts have found EAs should be used in these circumstances to determine whether "the new circumstance must present a seriously different picture of the environmental impact of the proposed project from what was previously envisioned."[69] If those different impacts are found, an EIS or Supplemental EIS is warranted. A federal district court considered whether a Supplemental EIS should have been prepared three years after conclusion of a multisale EIS for the Arctic region, instead of an EA. It found that the multisale EIS in that case had considered scenarios that included the new data addressed by the EA.[70] As this conclusion is fact-specific, it cannot be used to predict whether a court may reach the same decision regarding the Lease Sale 206 EA.

Categorical Exclusion for the Exploration Plan

The final environmental document to date for the Mississippi Canyon Block 252 oil well is a categorical exclusion for the exploration plan (EP) submitted by BP (EP CE). As discussed above, a categorical exclusion may be used by an agency when a category of actions has been determined to have no significant effect on the environment either individually or cumulatively.[71] The CEQ regulations provide that any procedures allowing CEs must also include "extraordinary circumstances" for when a CE would not apply.[72]

In 1986, MMS issued its list of categories of activities excluded from further NEPA review within the DOI Departmental Manual.[73] The list was revised in 2004. The Departmental Manual states that a CE does not apply if "the action qualifies as an exception under Appendix 2 of 516 DM." A copy of the MMS categorical exclusions and the exceptions is available as Appendix B to this report.[74]

MMS recorded its decision to invoke a categorical exclusion for the BP EP in a Categorical Exclusion Review (CER) with Analysis and included reviews for specific environmental harms. The CE process begins when a leaseholder (in this case BP) submits an exploration plan for MMS approval.[75] MMS reviews the plan to see if it fits any of the published categorical exclusions. The agency must also see whether any exceptions to those CEs apply. The decision must be documented. MMS uses forms that include check lists to document the CERs.

Under 43 U.S.C. § 1340(c), MMS is required to approve an exploration plan within 30 days of submission. The Secretary of the Interior, in testimony before the Senate Committee on Environment and Public Works, asked Congress to extend this deadline to 90 days.[76] This may suggest that a CE for the exploration plan was invoked because MMS had only 30 days to complete an environmental review. However, MMS prepares EAs (which require more analysis than a CE) for exploration plans that are not in the western or central Gulf of Mexico.[77]

Arguably, a 1976 U.S. Supreme Court decision supports the Secretary of the Interior's position that the deadline is too short.[78] In that decision, *Flint Ridge Development Co. v. Scenic Rivers Association of Oklahoma*, the Court held that a 30-day statutory deadline meant that the federal agency could not complete an EIS, and therefore NEPA did not require it. The Court noted that NEPA only required agencies to comply "to the fullest extent possible." The Court considered the fact that the 30-day deadline for the federal agency in *Flint Ridge* did not begin until the submission was completed. The Ninth

Circuit made a similar observation regarding the MMS 30- day deadline for exploration plan approval, finding some flexibility in the fact that a NEPA review could begin while the agency sought further information from an applicant.[79] However, that Ninth Circuit decision was withdrawn and vacated. *Flint Ridge* could be distinguished from the Mississippi Canyon EP in that the *Flint Ridge* plaintiffs sought to have a full EIS prepared, whereas for Mississippi Canyon the argument is that a CE was not enough, leaving open the possibility that an EA would suffice within the time frame.

According to MMS, CEs are available when "the impacts from the common operations are expected to be negligible to non-existent based upon general information gathered during past environmental analyses."[80] This means that based on its experience, MMS has found that those operations have at most negligible environmental impacts, and therefore, additional review is not required.

The CER documents from MMS do not indicate which CE was used. However, it seems likely that MMS invoked CE 15.4(C)(10), which is a CE that applies to exploration plans:

> Approval of an offshore lease or unit exploration development/ production plan or a Development Operation Coordination Document in the central or western Gulf of Mexico (30 CFR 250.2) except those proposing facilities: (1) In areas of high seismic risk or seismicity, relatively untested deep water, or remote areas, or (2) within the boundary of a proposed or established marine sanctuary, and/or within or near the boundary of a proposed or established wildlife refuge or areas of high biological sensitivity; or (3) in areas of hazardous natural bottom conditions; or (4) utilizing new or unusual technology.[81]

This CE appears to have been created in 1978 in guidelines for MMS's predecessor. It is the only CE that makes specific reference to an OCS area, or, in fact, contains any geographical restriction. It excludes two of the four stages of OCSLA actions from further NEPA review. It excludes both exploration plans in the western and central Gulf of Mexico and development and production plans in those areas. In contrast, the other CEs exclude what appear to be more general, administrative activities such as "issuance and modification of regulations, Orders, Standards, Notices to Lessees and Operators, Guidelines, and field rules" (C)(1); "approval of unitization agreements, pooling, or communitization agreements" (C)(4); and "approval of suspension of operations and suspensions of productions" (C)(6).

The CE for Gulf plans could be based on the OCSLA amendments of 1978.[82] Those amendments address certain NEPA requirements, giving exemptions to activities in the Gulf of Mexico. The relevant section of that law, Section 25 (codified at 43 U.S.C. § 1351(e)(1)), states that "at least once the Secretary shall declare the approval of a development and production plan in any area or region ... of the outer Continental Shelf, other than the Gulf of Mexico, to be a major Federal action."[83] While this language contains some ambiguity as to whether the "at least once" requirement must be used by the Secretary for all development plans or on a plan-by-plan basis, this suggests that approval of a development and production plan in the Gulf of Mexico might never need to be declared a *major federal action* (key words in NEPA to mandate preparation of an EIS).[84] No similar statutory reference to exploration plans exists, but this may be the foundation for excluding Gulf of Mexico plans from environmental analyses at certain stages.

BP discussed a Worst Case Scenario Response in its initial EP for Mississippi Canyon Block 252, indicating it considered a potential large-scale spill. The EP states that for exploratory wells A and B at Block 252, the worst case scenario would be a blowout at the exploratory stage, leading to a spill of 162,000 bbl[85] of crude oil a day.[86] The EP indicates that BP's regional oil spill response plan for a worst case scenario had been approved by MMS.[87] There is no indication that MMS conducted a NEPA review for approving the regional oil spill response plan (OSRP). This suggests that the methods for responding to a massive spill—such as in situ burning, or chemical dispersant—were not scrutinized for environmental impacts other than within the Multisale EIS and the Lease Sale 206 EA, and those documents considered much smaller leaks.

Despite referring to a blowout in the context of a response plan, no blowout scenario was included in the EP. This appears to be based on MMS policy for the area. Section 2.7 of the EP states: "A scenario for a potential blowout of the well from which BP would expect to have the highest volume of liquid hydrocarbons is not required for the operations proposed in this EP."[88] While MMS regulations require disclosure of a blowout scenario in EPs,[89] MMS provided an exception in a 2008 Notice to Lessees. The exception exempts OCS actions in the Gulf from blowout scenario requirements under certain conditions.[90] On June 18, 2010, this exception was revoked by another Notice. The 2010 Notice to Lessees expressly rescinds portions of the 2008 Notice; lessees not already having approved Applications for Permit to Drill are required to submit blowout scenarios and worst case discharge scenarios.[91]

Even if a blowout scenario was not prepared, it seems there would be significant amounts of oil released if the worst case scenario of a blowout occurred, oil at quantities greater than considered in the Multisale EIS or the Lease Sale 206 EA. However, because a CE was used for the EP, instead of an EIS or an EA, a review of the environmental impacts for a spill of this size was not conducted. They were not considered within the previous EISs or the EA either.

Adequacy of the Environmental Review of the Exploration Plan

No court has reviewed this categorical exclusion (516 DM 15.4(C)(10)), either to determine whether on its face it meets the standards for a CE—actions that either individually or cumulatively are found not to have a significant effect on the environment—or as it has been applied to any exploration or development plan. However, the Ninth Circuit considered whether MMS had properly used a different CE in California when approving lease suspensions. The court found the use of that CE was not adequately justified by MMS.[92] In that case 36 lease suspensions were sought nearly 30 years after a large oil spill near Santa Barbara. Under the OCLSA, MMS may suspend the term of a lease when a lessee is not able to begin production within the term of the lease, thus avoiding expiration of the lease.[93] The court found that there was no documentation of the CE, and that some exceptions to a CE could have applied that MMS did not consider. Accordingly, the court remanded the NEPA review to MMS.

It is possible a court may find that exceptions to CEs apply in the case of the EP CE, in which case the BP CE would be invalid. (The exceptions are included as part of Appendix B.) One exception that may apply addresses effects on ESA-listed species, meaning a CE could not be used if an action may "have adverse effects on species listed or proposed to be listed on the List of Endangered or Threatened Species, or have adverse effects on designated Critical Habitat for these species."[94] All of the environmental documents address the effects of an oil spill on listed species and the critical habitat of the Gulf Sturgeon. For example, the Lease Sale 206 EA states that over the 40-year term of the lease, 119 loggerhead, 10 leatherback, 1 hawksbill, 13 Kemp's ridley, and 38 green turtles will be killed.[95] It is possible a court could find that these effects could be considered "adverse." Another exception to the CEs addresses actions when the impacts cannot be known, perhaps due to the fact that the harm depends on the scale, timing, and location of an oil spill: "2.4 - Have highly uncertain and potentially significant environmental effects or involve unique or unknown environmental risks."

Even if the exceptions do not apply, MMS may be required to explain in the CE why they do not. Under accepted law, if there is substantial evidence that an exception *may* apply, an agency must document why an exception to a CE does *not* apply in order to support the CE.[96] While the CER documents specifically address exceptions regarding archeological resources and chemosynthetic communities, finding that neither exception applies, there is no explicit reference to any other exception.

The fact that this project was in the Gulf of Mexico made a difference in the type of review MMS conducted. Arguably, a heightened level of review might have reached a different conclusion as to the probable environmental impacts. If the EP had been in the Arctic, the CE for the Gulf plans would not apply and consideration of a blowout scenario also would have been required. Instead, EAs are used, which must sufficiently analyze the environmental impacts of the action. When the Ninth Circuit reviewed an EA for an exploration plan in the Beaufort Sea in the Arctic (hereinafter the Beaufort Sea EP), it found that an EA was not adequate.[97] The BP EP and the Beaufort Sea EP are factually similar in that they result from a nearly identical set of tiered NEPA analyses. The Beaufort Sea EP also came after an EIS for a 5-year plan (for 2002-2007), an EIS for the multiple lease sale, and a series of EAs for particular lease sales. Like the EP CE, endangered species were present at the planned drilling site. In the case of the Beaufort Sea EP the court indicated experts expressed concern that the exploration plan could have significant impacts on polar bears and whales. An EA-FONNSI was issued, meaning no new significant impacts were found. The court held that the Beaufort Sea EA did not indicate MMS took a "hard look" at the environmental impacts of the exploration plan since substantial questions remained regarding the harm to wildlife and the people in the area.[98] MMS revoked its approval of the EP.[99]

The EP CE, perhaps, could be justified as a fourth environmental review, implying that all of the environmental impacts had been considered in the earlier, more expansive documents. Technically, however, that is not how a CE is intended to be applied. The NEPA regulations do not provide that CEs may be used to exclude reviews on the basis that previous analyses already considered impacts. A CE is supposed to be invoked because the project would not have any impacts. NEPA allows tiered environmental reviews[100] when the impacts have already been considered.

Additionally, this rationale runs counter to the OCSLA staging justification used by courts as to why environmental reviews at the lease sale stage or earlier did not need to be in depth. Those courts held that a more intense environmental review was not needed at the early stages of the OCSLA because it would occur when actual drilling was imminent.[101] For example, the justification used by the District Court of Alaska for why a

100,000 bbl spill did not have to be analyzed may not apply when a CE is used at the exploration stage instead of a more in-depth analysis: "Congress has decided to allow key decisions having serious environmental consequences to be made at the exploration and production and development stages instead of requiring all decisions to be made at the pre-leasing and leasing stages."[102]

The Ninth Circuit's rationale for why a larger oil spill was not examined at the lease sale stage was that environmentally significant activities did not occur then. A more thorough review would be conducted later

> by purchasing a lease, lessees acquire no right to do anything more. Under the plain language of OCSLA, the purchase of a lease entails no right to proceed with full exploration, development, or production ... the lessee acquires only a priority in submitting plans to conduct those activities. If these plans, when ultimately submitted, are disapproved, no further exploration or development is permitted.[103]

If no review were conducted later, this justification may not seem to be supportable. The Ninth Circuit had suggested in another case that the exploration stage was where a spill analysis would be most appropriate.[104]

The U.S. Supreme Court also emphasized the importance of administrative review of the post-lease actions: "an OCS lease authorizes the holder to engage only in preliminary exploration; further administrative approval is required before full exploration or development may begin."[105] It could be argued that the EP CE is the inverse result of what is supposed to happen with tiered environmental reviews. Instead of the process described by the courts where the environmental review becomes more exacting as the drilling becomes imminent, the opposite has happened and no additional environmental analysis was conducted.

The Council on Environmental Quality Report on MMS and NEPA

In August 2010, CEQ issued a report on MMS/BOEMRE and its NEPA procedures for OCS drilling.[106] The report recommends that BOEMRE policies could be revised to reflect the purposes of NEPA more closely. Many of the recommendations address issues discussed within this report. For example, CEQ recommends that BOEMRE seek a statutory revision to OCSLA to extend the 30-day limit for considering EPs.

It also questions the propriety of allowing CEs for exploration and drilling and production plans. For one thing, it suggests that those CEs were developed before OCS activities were conducted at such depths, saying that there is a "progressively more complex environment" for deepwater operations than there was at the time of the CEs' creation. According to an MMS report, drilling at 1,000 feet began in the late 1970s, although drilling at 3,000 to 5,000 feet did not occur until the late 1990s.[107]

Another point the CEQ Report makes is that BOEMRE should review how it perceives "extraordinary circumstances." Under NEPA practice, finding an extraordinary circumstance indicates that a CE should not be used. The CEQ Report suggests that MMS was not properly evaluating those factors, stating that BOEMRE "intends to review its interpretation of the threshold requirement of 'extraordinary circumstances.'"[108]

Additionally, the CEQ Report points to the lack of consideration of massive oil spills in any of the environmental documents. While MMS found that a large spill had a low probability, the report still recommends that the agency consider reasonably foreseeable impacts associated with a catastrophic spill, despite the predicted low risk. The reasonably foreseeable impacts would include cumulative impacts. The cumulative impacts analysis for other Gulf projects should be different following the 4.1 million bbl spill.

On the same day as the CEQ Report was issued, the Department of the Interior announced it was not going to use certain CEs until it completed a review of all its CEs for OCS exploration and drilling activities.[109] The directive specifically prohibits using the CE that allows exclusion of exploration and development and production plans in the western and central Gulf (516 DM 15.4(C)(10)) if the project involves a subsea blowout preventer.

Congressional Attention to Categorical Exclusions

Congress has also focused on the use of CEs for exploration and development in the Gulf. Two House bills would eliminate the use of CEs for exploration, development, and production plans. H.R. 52 (Connolly) would require an EIS for those plans. It states that exploration, development, and production plans are "major federal actions significantly affecting the quality of the human environment." It would also eliminate the 30-day deadline for agency response to exploration plans.

H.R. 501 (Markey) is similar. It would require either an EA or an EIS for those three types of plans. Section 215(3) of H.R. 501 addresses development and production plans. Section 208(g) of H.R. 501 addresses exploration plans, extending the approval time for those plans from 30 to 90 days. In addition, it would require those plans to demonstrate the use of best available technology, including for blowout prevention. The bill would also put into law exceptional circumstances for when a plan can be rejected by the agency.

A third bill of the 112th Congress comes from the Senate. S. 203 (Begich) would prevent NEPA from applying to tests of technology for spill prevention, recovery, and mitigation in Arctic waters. It does this by declaring that such testing is not a major federal action. The bill also provides that the agency decision regarding this exclusion is exempt from judicial review.

In the 111th Congress, three bills were proposed by the House to restrict the use of CEs. H.R. 3534 (Rahall), the Consolidated Land, Energy, and Aquatic Resources Act of 2010 (CLEAR), would have required BOEMRE approval of exploration, development, and production plans to include either an EA or an EIS, effectively eliminating the CE for those plans.[110] CLEAR would have amended the requirement for approval of EPs from 30 days to 90 days.[111] H.R. 5506 (Connolly) would have gone further than H.R. 3534 by requiring an EIS (and not an EA) for those plans' approval. It would have removed the 30-day deadline. H.R. 5572 (Buchanan) states that the Department of the Interior could not exempt any OCS activity from legal requirements.

CONCLUSION

The exploration well blowout in the Gulf of Mexico appears to have major, and potentially catastrophic, environmental effects. None of the environmental reviews under NEPA considered an oil spill of this magnitude. However, it is unclear whether that violates NEPA, which has no requirement for analysis of a worst case scenario. The use of a categorical exclusion at the exploration stage appears to be procedurally correct under MMS policy, provided there were no applicable exceptions. However, the legality of that CE has never been examined by a court, which could consider whether an exemption from a NEPA review at the point where drilling is imminent violates NEPA either because extraordinary circumstances existed in the case, or because, in general, the tiered OCSLA process envisioned more scrutiny of environmental consequences after the lease stage. BOEMRE has indicated it will take a closer look at its CEs and how they are applied.

APPENDIX A. BACKGROUND FACTS

Source: MMS, Gulf of Mexico OCS Oil and Gas Lease Sales: Final Environmental Impact Statement (MMS 2007-018) (April 2007).

Duration of Gulf of Mexico OCS Oil and Gas Activity—40 years (p. 4-5):

- Exploratory drilling activity can take place over an 8-year period, beginning within one year after the lease sale.
- Development activity takes place over a 39-year period, beginning with the installation of the first production platform and ending with the drilling of the last development wells.
- Production of oil and gas begins by the third year after the lease sale and continues through the 40th year.
- Final abandonment and removal activities occur in the 40th year.

Amount of oil and gas expected to be produced (p. 4-5):

Western Gulf:
- Oil = 242-423 million bbl
- Gas = 1.644-2,647 trillion cubic feet

Central Gulf:

- Oil = 776-1,292 million bbl
- Gas = 3.236-5.229 trillion cubic feet

Amount of seismic testing expected (p. 4-7):

- Western Gulf—400-800 blocks
- Central Gulf—1,000-2,000 blocks

Number of exploration and delineation wells from this lease sale expected during 40-year period (p. 4-10):

- Western Gulf—42-66
- Central Gulf—65-96

Number of development wells from this lease sale anticipated during 40-year period (p. 4-13):

- Western Gulf—155-221
- Central Gulf—330-468

Number of loss of well control events (including blowouts) from this lease sale estimated during 40-year period (p. 4-250):

- Western Gulf—1-2
- Central Gulf—2-3

Number of offshore oil spills estimated from this lease sale during 40-year period (p. 4-241):

- Western Gulf—800-1,500 spills/ 400-21,000 bbl of oil
- Central Gulf—2,700-4,500 spills/ 5,500-26,500 bbl of oil

Natural seeps of oil, annual average (1990-1999) (Table 4-12):

- Western Gulf—490,000 bbl
- Eastern Gulf[112]—490,000 bbl

Probability of spills greater than 10,000 bbl occurring (Table 4-15):

Western Gulf:

- Mean number of spills from all sources (facilities, pipelines, shuttle tankers)—3
- Probability of occurrence—92%-96%

Central Gulf:

- Mean number of spills from all sources (facilities, pipelines, shuttle tankers)—9
- Probability of occurrence—99+%

Historical data on offshore spills greater than 10,000 bbl in the Gulf of Mexico from 1985-1999 (Table 4-16):

- From pipelines: 2 spills; 30,000 bbl spilled in total
- From platforms: 0 spills

APPENDIX B. MMS CATEGORICAL EXCLUSIONS

Source: 516 DM

15.4 Categorical Exclusions. In addition to the actions listed in the Departmental categorical exclusions outlined in Appendix 1 of 516 DM 2, many of which the MMS also performs, the following MMS actions are designated categorical exclusions unless the action qualifies as an exception under Appendix 2 of 516 DM 2:

A. General

1. Inventory, data, and information collection, including the conduct of environmental monitoring and nondestructive research programs.
2. Actions for which MMS has concurrence or co-approval with another Bureau if the action is a categorical exclusion for that Bureau.

B. Internal Program Initiatives

1. All resource evaluation activities including surveying, mapping, and geophysical surveying which do not use solid or liquid explosives.
2. Collection of geologic data and samples including geologic, paleontologic, mineralogic, geochemical, and geophysical investigations which does not involve drilling beyond 50 feet of consolidated rock or beyond 300 feet of unconsolidated rock, including contracts therefor.
3. Acquisition of existing geological or geophysical data from otherwise private exploration ventures.
4. Well logging, digital modeling. inventory of existing wells, and installation of recording devices in wells.
5. Establishment and installation of any research/monitoring devices.

6. Test or exploration drilling and downhole testing included in a project previously subject to the NEPA process.
7. Insignificant revisions to the approved 5-year leasing program.
8. Prelease planning steps such as the Call for Information and Area Identification.

C. Permit and Regulatory Functions

1) Issuance and modification of regulations, Orders, Standards, Notices to Lessees and Operators. Guidelines and field rules for which the impacts are limited to administrative, economic, or technological effects and the environmental impacts are minimal.
2) Approval of production measurement methods, facilities, and procedures.
3) Approval of off-lease storage in existing facilities.
4) Approval of unitization agreements, pooling, or communitization agreements.
5) Approval of commingling of production.
6) Approval of suspensions of operations and suspensions of production.
7) Approval of lease consolidation applications, lease assignments or transfers, operating rights, operating agreements, lease extensions, lease relinquishments, and bond terminations.
8) Administration decisions and actions and record keeping such as:

a. Approval of applications for pricing determinations under the Natural Gas Policy Act.
b. Approval of underground gas storage agreements from a presently or formerly productive reservoir.
c. Issuance of paying well determinations and participating area approvals.
d. Issuance of drainage determinations.

9) Approval of offshore geological and geophysical mineral exploration activities, except when the proposed activity includes the drilling of deep stratigraphic test holes or uses solid or liquid explosives.
10) Approval of an offshore lease or unit exploration. development/ production plan or a Development Operation Coordination Document in the central or western Gulf of Mexico (30 CFR 250.2) except those

proposing facilities: (1) In areas of high seismic risk or seismicity, relatively untested deep water, or remote areas, or (2) within the boundary of a proposed or established marine sanctuary, and/or within or near the boundary of a proposed or established wildlife refuge or areas of high biological sensitivity; or (3) in areas of hazardous natural bottom conditions; or (4) utilizing new or unusual technology.

11) Approval of minor revisions of or minor variances from activities described in an approved offshore exploration or development/ production plan, including pipeline applications.
12) Approval of an Application for Permit to Drill (APD) an offshore oil and gas exploration or development well, when said well and appropriate mitigation measures are described in an approved exploration plan, development plan, production plan, or Development Operations Coordination Document.
13) Preliminary activities conducted on a lease prior to approval of an exploration or development/production plan or a Development Operations Coordination Plan. These are activities such as geological, geophysical, and other surveys necessary to develop a comprehensive exploration plan, development/production plan, or Development Operations Coordination Plan.
14) Approval of Sundry Notices and Reports on Wells.
15) Rights-of-ways, easements, temporary use permits, and any revisions thereto that do not result in a new pipeline corridor to shore.

D. Royalty Functions

All functions of the Associate Director for Royalty Management including, but not limited to, such activities as: approval of royalty payment procedures, including royalty oil contracts; and determinations concerning royalty quantities and values, such as audits, royalty reductions, collection procedures, reporting procedures, and any actions taken with regard to royalty collections (including similar actions relating to net profit and windfall profit taxes).

Exceptions to Categorical Exclusions

516 DM 2 Appendix 2

The following exceptions apply to individual actions within categorical exclusions (CX). Environmental documents must be prepared for actions which may:

2.1. Have significant adverse effects on public health or safety

2.2. Have adverse effects on such unique geographic characteristics as historic or cultural resources, park, recreation or refuge lands, wilderness areas, wild or scenic rivers, sole or principal drinking water aquifers, prime farmlands, wetlands, floodplains or ecologically significant or critical areas, including those listed on the Department's National Register of Natural Landmarks.

2.3. Have highly controversial environmental effects

2.4. Have highly uncertain and potentially significant environmental effects or involve unique or unknown environmental risks.

2.5 Establish a precedent for future action or represent a decision in principle about future actions with potentially significant environmental effects.

2.6 Be directly related to other actions with individually insignificant but cumulatively significant environmental effects.

2.7 Have adverse effects on properties listed or eligible for listing on the National Register of Historic Places.

2.8 Have adverse effects on species listed or proposed to be listed on the List of Endangered or Threatened Species, or have adverse effects on designated Critical Habitat for these species.

2.9 Require compliance with Executive Order 11988 (Floodplain Management), Executive Order 11990 (Protection of Wetlands), or the Fish and Wildlife Coordination Act.

2.10 Threaten to violate a Federal, State, local or tribal law or requirement imposed for the protection of the environment.

Appendix C. The MMS Categorical Exclusion Review Process

Source: Minerals Management Service

A more detailed discussion of Categorical Exclusion Reviews (CERs) and the NEPA process for review of Industry Submittals follows:

Descriptions for the "NEPA Determination Type" field under FastFacts?

The **NEPA Determination Type** is identified in a **NEPA Determination Type Review** at the very beginning of the Categorical Exclusion Review (CER) process and it establishes the level of NEPA analyses that should be conducted for a particular proposal. Based upon details about the proposed action, policy, and programmatic NEPA analyses, the NEPA Determination Type could be one of the following:

- **CER with no further Analyses**; i.e., finalize the CER with no additional reviews/conditions of approval (COAs)
- **CER with Analyses**; i.e., continue processing/review under a CER with the applicable protected resource reviews assigned; or
- **Prepare an Environmental Assessment (EA)** or an **Environmental Impact Statement (EIS)**

When it's a categorical exclusion *[review]* with analysis, what kind of analysis is it?

Though not a document in the same sense as an EA or EIS, the CERs ultimately consist of a series of procedures and activity-specific reviews that are conducted and compiled so that MMS is assured that the activity can remain Categorically-Excluded from review under an EA or EIS. The CERs are completely digital and therefore, stored within TIMS as part of the Administrative Record along with all documentation related to the proposed action (in this case, the EPs on MC252) and all associated surveys/reports. Even though central and western Gulf of Mexico drilling and production activities are Categorically-Excluded from any additional review, MMS GOMR prepares CERs on each proposal to ensure that "extraordinary circumstances" do not exist and to ensure that the agency has the ability to consider the best-available data/technology on a case-by-case basis, which in turn affords real-time adaptive management of assessment triggers, reviewing standards, and conditions of approval/mitigation.

When an exception criteria *[now called an "extraordinary circumstance"]* is identified during analysis, what exactly does that mean?

An "extraordinary circumstance" (EC) is an identified condition that could exist within an action that could be Categorically-Excluded from additional NEPA analysis. MMS has 12 ECs identified under 43 CFR § 46.215 (see last page) that it reviews the proposed action against to determine if any of the activities proposed by the operator may "trigger" or cause an EC condition to occur. If an EC is identified for a proposal that cannot be avoided (either in the NEPA Determination Review or protected resource reviews), an EA must be prepared. The EA, in turn, could result in a Finding of Significant Impact (FOSI), which would then require preparation of an EIS.

Detailed Discussion

The DOI and MMS can categorically-exclude (CatEx) certain actions from further NEPA analyses if it is determined that they would not have a significant effect on the environment (individually or cumulatively). The MMS CatEx list is currently found in the Departmental Manual (see 516DM15.4) and it not only contains administrative actions, but also common OCS operations. The impacts from the common operations are expected to be negligible to nonexistent based upon general information gathered during past environmental analyses about the type of activities, the area of the proposed action, programmatic NEPA analyses (Program/Sale EISs/EAs), and the past actions as conducted and observed. The specific MMS operational CatEx corresponding to the submittal of an Exploration Plan (EP) is below:

Since the CatEx list is based upon general/historical information and generalized/programmatic information, GOMR goes the extra step and prepares a Categorical Exclusion Review (CER) for each action to determine if site-specific details of the proposed operations will not result in an Extraordinary Circumstance (EC), which would require that an EA be prepared. Coordinators in GOMR's Environmental Compliance Section (ECS) follow a set of submittal-specific SOPs and utilize a detailed GIS application, PMT GIS, that addresses the activity being proposed and compares information with established business rules/EC triggers based upon the specific resources to determine if the action should be reviewed under an Environmental Assessment (EA) and which, if any, protected resource reviews should be assigned for the CER/EA.

> MMS Categorical Exclusion
>
> under 516 DM 15.4
>
> C. Permit and Regulatory Functions.
>
> 10. Approval of an offshore lease or unit exploration plan (EP), development/production plan (DPP), or a Development Operation Coordination Document (DOCD) in the central or western Gulf of Mexico (30CFR250.2) except those proposing facilities:
>
> In areas of high seismic risk or seismicity, relatively untested deep water, or remote areas, or
>
> within the boundary of a proposed or established marine sanctuary, and/or within or near the boundary of a proposed or established wildlife refuge or areas of high biological sensitivity; or
>
> in areas of hazardous natural bottom conditions; or
>
> utilizing new or unusual technology.

When the NEPA coordinator is finished following the SOP, their NEPA Determination Review and associated *PMT GIS* report will identify the Initial **NEPA Determination Type**, which could include the following:

- **CER with no further Analyses**; i.e., finalize the CER with no additional reviews/conditions of approval (COAs)
- **CER with Analyses**; i.e., continue processing/review under a CER with the applicable protected resource reviews assigned; or
- **Prepare an EA/EIS**.

All of the applicable information related to the NEPA Determination and/or any resource reviews is recorded in the Technical Information Management System (TIMS) and the *PMT GIS* Report is attached in the respective NEPA Determination Type Review. A copy of the plan/application is provided to the NEPA Decision Maker for a concurrence review of the TIMS information and *PMT GIS* Report. If the Decision Maker concurs with the coordinator, NEPA Determination Type is also recorded in TIMS for the specified submittal. If the Chief has any questions about the NEPA determination type, the information will be reviewed together until a decision can be made.

Once concurrence is given, all necessary reviews are assigned to resource specialists (i.e., marine archaeologists, benthic biologists, meteorologists, oceanographers, etc.) and tracked in TIMS. The SME informs the NEPA Coordinator if additional information or a clarification is required and/or when the review is completed. The completed review will note one of the following:

- The proposed action will have no impact on the protected resource.
- The proposed action requires a condition of approval (COA) to ensure that an extraordinary circumstance (EC) will not be met/triggered.
- The proposed action may cause a significant impact and an EA is recommended.

When all of the applicable reviews are completed, the SME's conclusions, and any suggested conditions of approval/mitigation (if necessary), are recorded in TIMS and compiled by the NEPA Coordinator for presentation to the NEPA Decision Maker. The NEPA decision maker will review the findings and choose from the following Final NEPA Actions:

- **CER Approved;** (Action can be CatExed with no COAs—an EC will not be triggered/occur); or
- **CER Approved;** (Action can be CatExed with COAs—to ensure that an EC will not be triggered/occur);
- **Prepare an EA/EIS.**

43 CFR § 46.215 Categorical Exclusions: Extraordinary circumstances

Extraordinary circumstances (see paragraph 46.205(c)) exist for individual actions within categorical exclusions that may meet any of the criteria listed in paragraphs (a) through (l) of this section. Applicability of extraordinary circumstances to categorical exclusions is determined by the Responsible Official.

- Have significant impacts on public health or safety.
- Have significant impacts on such natural resources and unique geographic characteristics as historic or cultural resources; park, recreation or refuge lands; wilderness areas; wild or scenic rivers; national natural landmarks; sole or principal drinking water aquifers; prime farmlands; wetlands (EO 11990); floodplains (EO 11988); national monuments; migratory birds; and other ecologically significant or critical areas.
- Have highly controversial environmental effects or involve unresolved conflicts concerning alternative uses of available resources [NEPA section 102(2)(E)].
- Have highly uncertain and potentially significant environmental effects or involve unique or unknown environmental risks.
- Establish a precedent for future action or represent a decision in principle about future actions with potentially significant environmental effects.
- Have a direct relationship to other actions with individually insignificant but cumulatively significant environmental effects.
- Have significant impacts on properties listed, or eligible for listing, on the National Register of Historic Places as determined by the bureau.
- Have significant impacts on species listed, or proposed to be listed, on the List of Endangered or Threatened Species or have significant impacts on designated Critical Habitat for these species.
- Violate a Federal law, or a State, local, or tribal law or requirement imposed for the protection of the environment.
- Have a disproportionately high and adverse effect on low income or minority populations (EO 12898).
- Limit access to and ceremonial use of Indian sacred sites on Federal lands by Indian religious practitioners or significantly adversely affect the physical integrity of such sacred sites (EO 13007).

Contribute to the introduction, continued existence, or spread of noxious weeds or non-native invasive species known to occur in the area or actions that may promote the introduction, growth, or expansion of the range of such species (Federal Noxious Weed Control Act and EO 13112).Contribute to the introduction, continued existence, or spread of noxious weeds or nonnative invasive species known to occur in the area or actions that may promote the introduction, growth, or expansion of the range of such species (Federal Noxious Weed Control Act and EO 13112).

End Notes

[1] Department of the Interior Press Release, *Flow Rate Group Provides Preliminary Best Estimate Of Oil Flowing from BP Oil Well* (May 27, 2010). Available at http://www.doi.gov/news/pressreleases/Flow-Rate-Group-ProvidesPreliminary-Best-Estimate-Of-Oil-Flowing-from-BP-Oil-Well.cfm. Estimates were revised from 5,000 bbl a day to 62,000 bbl a day, decreasing toward the end of the leak to 53,000 bbl. An estimated total of 4.1 million bbl were released into the Gulf (4.9 million bbl leaked, but 800,000 bbl captured before it leaked into the Gulf). See Official Site of the Deepwater Horizon Unified Command, at http://www.deepwaterhorizonresponse.com/ go/doc/2931/840475.

[2] 43 U.S.C. §§ 1331 – 1356a. For analysis of the OCSLA, see CRS Report RL33404, *Offshore Oil and Gas Development: Legal Framework*, by Adam Vann.

[3] 16 U.S.C. §§ 1531 – 1544. See Section 1536 for consultation obligations of federal agencies.

[4] 16 U.S.C. §§ 1361 – 1384. See Section 1371(a)(5)(A) for incidental takes of marine mammals.

[5] 16 U.S.C. §§ 1451 – 1456. See Section 1456(c) for federal actions affecting coasts.

[6] Secretarial Order 3299 (May 19, 2010), available at http://www.doi.gov/deep waterhorizon/loader.cfm?csModule= security/getfile&PageID=32475.

[7] 42 U.S.C. § 4332. The act also requires consideration of "(iv) the relationship between local short-term uses of man's environment and the maintenance and enhancement of long-term productivity, and (v) any irreversible and irretrievable commitments of resources."

[8] As referenced above, MMS was reorganized, effective May 19, 2010. Secretary of the Interior Order No. 3299.

[9] Kleppe v. Sierra Club, 427 U.S. 390, 410 n.21 (1976) (the role of a court is to ensure than an agency took a "hard look" at the environmental consequences); National Audubon Society v. Department of the Navy, 422 F.3d 174, 185 (4th Cir. 2005) (a hard look "encompasses a thorough investigation into the environmental impacts of an agency's action and a candid acknowledgement of the risks that those impacts entail"); Idaho Conservation League v. Mumma, 956 F.2d 1508, 1519 (9th Cir. 1992) (courts review whether an EIS "contains a reasonably thorough discussion of the significant aspects of the probable environmental consequences"); Natural Resources Defense Council, Inc. v. Morton, 458 F.2d 827, at 838 (D.C. Cir. 1972) (an environmental review complies with NEPA "so long as the officials and agencies have taken the 'hard look' at environmental consequences mandated by Congress").

[10] 40 C.F.R. § 1506.6. See California v. Block, 590 F.2d 753 (9th Cir. 1982) (informed public participation is a goal of NEPA).

[11] 40 C.F.R. § 1508.4.

[12] California v. Norton, 311 F.3d 1162 (9th Cir. 2002).

[13] 40 C.F.R. part 1500.

[14] 516 DM 1-15, available at http://206.131.241.18/app_DM/index.cfm? fuseaction=search DM&keyword= 516%20DM%201-15.

[15] 516 DM 15.

[16] A copy of this is available at http://www.boemre.gov/PDFs/5BOEMRE_Leasing101.pdf.

[17] 43 U.S.C. § 1344—5-year plan; 43 U.S.C. § 1337(a)—lease sale; 43 U.S.C. § 1340—exploration plan; and 43 U.S.C. § 1351—development and production.

[18] 40 C.F.R. § 1500.4: "Agencies shall reduce excessive paperwork by ... (i) using program, policy, or plan environmental impact statements and tiering from statements of broad scope to those of narrower scope, to eliminate repetitive discussions of the same issues."

[19] MMS, Geological and Geophysical Exploration for Mineral Resources on the Gulf of Mexico Outer Continental Shelf - Final Programmatic Environmental Assessment, MMS 2004-054 (July 2004). 69 Fed. Reg. 45851 (July 30, 2004).

[20] MMS, Structure-Removal Operations on the Gulf of Mexico Outer Continental Shelf Programmatic Environmental Assessment, MMS 2005-013 (February 2005). 70 Fed. Reg. 9965 (March 1, 2005).

[21] 71 Fed. Reg. 50457, 50458 (Aug. 25, 2006).

[22] 72 Fed. Reg. 24326 (May 2, 2007).

[23] Center for Biological Diversity v. U.S. Department of the Interior, 563 F.3d 466 (D.C. Cir. 2009).

[24] 75 Fed. Reg. 16833 (April 2, 2010). Available at http://www.boemre.gov/5-year/2007-2012FEIS.htm. On that same date, MMS announced intent to prepare an EIS for the next 5-year plan, from 2012 to 2017. 75 Fed. Reg. 16828 (April 2, 2010).

[25] MMS, Outer Continental Shelf Oil & Gas Leasing Program: 2007-2012, Final Environmental Impact Statement MMS 2007-003 (hereinafter 5-year Plan EIS). Available at http://www.boemre.gov/5-year/2007-2012FEIS.htm.

[26] 5-year Plan EIS, Ch. IV, pp. IV-28 – IV-31.

[27] 5-year Plan EIS, Ch. IV, pp. IV-29.

[28] 5-year Plan EIS, Table IV-4.

[29] 5-year Plan EIS, Ch. IV, pp. IV-33 – IV-102.

[30] 43 U.S.C. § 1349(c)(1) gives this court exclusive jurisdiction for review of OCSLA leasing program approvals. This exclusive jurisdiction has been found to apply only to OCSLA claims and not to NEPA claims related to OCSLA approval. See Get Out Oil, Inc. v. Andrus, 477 F. Supp. 40, 42 (C.D. Cal. 1979).

[31] Center for Biological Diversity v. U.S. Department of the Interior, 563 F.3d 466 (D.C. Cir. 2009).

[32] Center for Biological Diversity v. U.S. Department of the Interior, 563 F.3d 466, 480 (D.C. Cir. 2009).

[33] Mobil Oil Corp. v. FTC, 562 F.2d 170, 173 (2d Cir. 1977) (finding that certain adjudicatory proceedings were not subject to NEPA because the final order was still speculative).

[34] Wyoming Outdoor Council v. U.S. Forest Service, 165 F.3d 43 (D.C. Cir. 1999). See also Conner v. Burford, 848 F.2d 1441 (9th Cir. 1988) (holding that no EIS was required where lease was issued forbidding surface occupancy (and therefore oil drilling)).

[35] MMS, Gulf of Mexico OCS Oil and Gas Lease Sales: 2007-2012; Western Planning Area Sales 204, 207, 210, 215, and 218; Central Planning Area Sales 205, 206, 208, 213, 216, and 222; Final Environmental Impact Statement, MMS 2007-018 (April 2007) (hereinafter Multisale EIS). Available at http://www.gomr.boemre.gov/homepg/regulate/ environ/nepa/ nepaprocess.html.

[36] Approximately 5.8 million acres located in the southeastern part of the Central Planning Area (CPA) are not included in the lease sale area, despite being opened to leasing by the Gulf of Mexico Energy Security Act of 2006 after years of leasing moratoria in appropriations acts. 71 Fed. Reg. 35258 (June 27, 2007).

[37] Multisale EIS, p. 4-232.

[38] Multisale EIS, p. 4-228. Ninety-four percent of all spills in the Gulf were found to be less than 1 bbl, contributing 5% by volume to all spills. Multisale EIS, p. 4-235.

[39] Multisale EIS, p. 4-233.

[40] Multisale EIS, p. 4-75.

[41] Id.

[42] Multisale EIS, p. 4-232.

[43] Multisale EIS, p. 4-243.

[44] Multisale EIS, p. 4-239.

[45] Multisale EIS, p. 4-260.

[46] Multisale EIS, p. 4-275.

[47] 42 U.S.C. § 4332(2)(C)(ii).

[48] Kleppe v. Sierra Club, 427 U.S. 390, 410 (1976).

[49] Idaho Conservation League v. Mumma, 956 F.2d 1508, 1519 (9th Cir. 1992).

[50] County of Suffolk v. Dept. of the Interior, 562 F.2d 1368 (2d Cir. 1977) (holding that an EIS for OCS drilling proposal was not inadequate).

[51] The Multisale EIS for the Gulf reviewed 11 lease sales.

[52] Tribal Village of Akutan v. Hodel, 869 F.2d 1185 (9th Cir. 1988).

[53] Id. at 1192.

[54] Id.

[55] Secretary of the Interior v. California, 464 U.S. 312 (1984) (holding that an adverse effects analysis under the CZMA was not required at the lease sale stage).

[56] Secretary of the Interior v. California, 464 U.S. 312, 317 (1984).

[57] Village of False Pass v. Watt, 565 F. Supp. 1123 (D.C. Alaska 1983), *aff'd sub nom.*, Village of False Pass v. Clark, 733 F.2d 605 (9th Cir. 1984).

[58] Village of False Pass v. Watt, 565 F. Supp. 1123, 1147 (D.C. Alaska 1983).

[59] Village of False Pass v. Clark, 733 F.2d 605, 616 (9th Cir. 1984).

[60] Robertson v. Methow Valley Citizens Council, 490 U.S. 332 (1989); Sierra Club v. Sigler, 695 F.2d 957 (5th Cir. 1983) (holding the Corps of Engineers did not have to consider the worst case scenario of a tanker losing all of its cargo when considering the impacts of a dredging permit basing its holding in significant part on a now-revised CEQ regulation).

[61] 40 C.F.R. § 1502.22(b) (revised in 1986). The current regulations require an agency to prepare a statement regarding unavailable information that includes *reasonably foreseeable significant adverse impacts*, which is defined as including "impacts which have catastrophic consequences, even if their probability of occurrence is low, provided that the analysis of the impacts is supported by credible scientific evidence."

[62] Grazing Fields Farm v. Goldschmidt, 626 F.2d 1068, 1074 (1st Cir.1980).

[63] MMS, Proposed Gulf of Mexico OCS Oil and Gas Lease Sale 206; Central Planning Area; Environmental Assessment, MMS 2007-059 (Oct. 2007) (hereinafter Lease 206 EA). Available at http://www.gomr.boemre.gov/PDFs/ 2007/2007-059.pdf.

[64] EA FONNSI, introduction.

[65] EA FONNSI, p. 1.

[66] Lease 206 EA, p. 14.

[67] Lease 206 EA, p. 18.

[68] Lease 206 EA, p. 24.

[69] Sierra Club v. Froehlke, 816 F.2d 205, 210 (5th Cir. 1987); North Slope Borough v. Minerals Management Service, No. 3:07-cv-0045-RRB, *2 (D. Alaska April 12, 2007); Oregon Natural Resources Council v. Devlin, 776 F. Supp. 1440, 1449 (D. Ore. 1991).

[70] North Slope Borough v. Minerals Management Service, No. 3:07-cv-0045-RRB, *2 (D. Alaska April 12, 2007).

[71] 40 C.F.R. § 1508.4.

[72] 40 C.F.R. § 1508.4.

[73] The *Federal Register* notice announcing these CEs refers to them as being functions transferred from the Geological Survey (GS) and the Bureau of Land Management (BLM) to the MMS upon its establishment. 51 Fed. Reg. 1855 (Jan. 15, 1986).

[74] It is also available at http://206.131.241.18/app_DM/act_getfiles.cfm?relnum=3625.

[75] In practice, multiple versions of an EP may be submitted, such as an initial EP, or an amended EP. The CER for BP shows separate reviews were conducted for archaeological resources, chemosynthetic communities, and what was referred to as a NEPA (1) determination. Copies of these documents (either the I-EP (Initial Exploration Plan) with a NEPA final date of April 15, 2009, or the A-EP (Amended Exploration Plan) with a NEPA final date of April 3, 2009) are available from the author.

[76] Senate EPW Hearing, *Federal Response to the Recent Oil Spill in the Gulf of Mexico* (May 18, 2010).

[77] For example, MMS prepared an EA for an exploration plan in the Arctic that was rejected by a court and ultimately withdrawn by the agency. *Alaska Wilderness League v. Kempthorne*, 548 F.3d 815 (9th Cir. 2008), *withdrawn and vacated as moot* 559 F.3d 916 (9th Cir. 2009), *and superseded sub nom. Alaska Wilderness League v. Salazar*, 571 F.3d 859 (9th Cir. 2009) (upon rescission by MMS of the EP approval).

[78] Flint Ridge Development Co. v. Scenic Rivers Association of Oklahoma, 426 U.S. 776 (1976) (holding that 30 days after filing was not enough time to complete a NEPA review and therefore one was not required).

[79] Alaska Wilderness League v. Kempthorne, 548 F.3d 815 (9th Cir. 2008), *withdrawn and vacated as moot*, 559 F.3d 916 (9th Cir. 2009), *and superseded sub nom.* Alaska Wilderness League v. Salazar, 571 F.3d 859 (9th Cir. 2009) (upon rescission by MMS of the EP approval).

[80] MMS, Discussion of Categorical Exclusion Reviews (undated). A copy of this document is included as Appendix C.

[81] 516 DM 15.4(C)(10).

[82] P.L. 95-372, § 25, 92 Stat. 659 (1978).

[83] The legislative history for this amendment does not discuss why the Gulf was excepted from this provision. However, the Conference Report suggests that the exception was for efficiency: "to limit bureaucratic redtape and otherwise minimize delays in the search for and production of oil and gas" which had been "going on for a number of years" in the Gulf. H. Rep. 95-1474, at 115 (Aug. 10, 1978). The Gulf exclusion was within the Senate language, while the House had no exclusion. Id.

[84] As referenced earlier in this report, MMS policy is that development and production plans undergo a full EIS. However, the MMS allows a CE for development and production plans in the western and central areas of the Gulf of Mexico, essentially exempting those plans from that policy. Because it was still in exploration, the Mississippi Canyon Block had not reached the stage of development and production. It is not known whether MMS would have issued a CE for that plan as well.

[85] The EP does not indicate the measure of the spill (either in bbl or gallons). The data given here are based on the fact that the Regional Oil Spill Response Plan (OSRP) by BP reports the worst case scenario for an exploratory well in barrels per day. See Regional Oil Spill Response Plan, Appx. H (Rev. 6/30/09), available at http://info.publicintelligence.net/

BPGoMspillresponseplan.pdf. The data in the EP for worst case discharge do not match that in the OSRP. The OSRP estimates a 250,000 bbl per day spill from MC 462 as the worst case discharge from an exploratory well. It does not have any reference to MC 252. See OSRP Quick Guide Table 1 for a list of production facilities. The EP estimates the daily volume as 300,000 bbl for MC 727, and 162,000 bbl for MC 252. MC 727 is not referenced in OSRP Table 1.

[86] EP, § 7.1.

[87] EP, § 7.1.

[88] BP, Initial Exploration Plan, Mississippi Canyon Block 252, OCS-G 32306 (received by MMS Feb. 23, 2009) (hereinafter EP). A copy of this document is available from the author.

[89] 30 C.F.R. § 250.213(g). A lawsuit challenging the DOI's issuance of the Notice was filed in federal court. Gulf Restoration Network v. Salazar, No. 2:10-cv-01497 (E.D. La. *filed* May 18, 2010).

[90] MMS, Notice to Lessee 2008-G04 (May 1, 2008). Available at http://www.gomr.mms.gov/homepg/regulate/regs/ ntls/2008NTLs/08-g04.pdf. Under this Notice, a blowout scenario is only required for OCS drilling if 1) Florida is an affected state; 2) the activity occurs within protective zones of Flower Garden Banks or Stetson Bank; 3) activity includes installation of surface facility at greater than 400 meters; 4) initial Development Operations Coordination Documents (DOCDs) and supplemental DOCDs with new multiwell structures if either Texas or Louisiana is an affected state; and 5) initial EPs if Texas is an affected state. Notice 2008 G-4, p. 6.

[91] MMS, Notice to Lessee No. 2010-N06. Available at http://www.doi.gov/deepwaterhorizon/loader.cfm?csModule= security/getfile&PageID=35724.

[92] California v. Norton, 311 F.3d 1162 (9th Cir. 2002).

[93] 43 U.S.C. § 1334(a)(1).

[94] 516 DM Appx. 2, exception 2.8.

[95] Lease 206 EA, p. 41.

[96] Jones v. Gordon, 792 F.2d 821 (9th Cir. 1986).

[97] Alaska Wilderness League v. Kempthorne, 548 F.3d 815 (9th Cir. 2008), *vacated as moot*, 571 F.3d 859 (9th Cir. 2009) (upon rescission by MMS of the EP approval).

[98] Alaska Wilderness League v. Kempthorne, 548 F.3d 815, 825 (9th Cir. 2008), *vacated as moot*, 571 F.3d 859 (9th Cir. 2009).

[99] See Alaska Wilderness League v. Kempthorne, 571 F.3d 859 (9th Cir. 2009).

[100] 40 C.F.R. § 1500.4(i).

[101] Center for Biological Diversity v. U.S. Department of the Interior, 563 F.3d 466, 480 (D.C. Cir. 2009).

[102] Village of False Pass v. Watt, 565 F. Supp. 1123 (D.C. Alaska 1983).

[103] Village of False Pass v. Clark, 733 F.2d 605, 608 (9th Cir. 1984).

[104] Tribal Village of Akutan v. Hodel, 869 F.2d 1185, 1192 (9th Cir. 1988).

[105] Secretary of the Interior v. California, 464 U.S. 312, 321 (1984).

[106] CEQ, Report Regarding the Minerals Management Service's National Environmental Policy Act Policies, Practices, and Procedures as They Relate to Outer Continental Shelf Oil and Gas Exploration and Development (Aug. 16, 2010) (hereinafter CEQ Report). Available at http://www.whitehouse.gov/sites/default/files/microsites/ceq/20100816-ceqmms-ocs-nepa.pdf.

[107] MMS, Deepwater Gulf of Mexico 2008: America's Offshore Energy Future, Appx. A (CS Report MMS 2008-013).

[108] CEQ Report, at 31.

[109] BOEMRE Directive, Use of Categorical Exclusions in Gulf of Mexico Region (Aug. 16, 2010). Available at http://www.doi.gov/news/pressreleases/loader.cfm?csModule =security/getfile&PageID=42011.

[110] H.R. 3534, § 208(f).

[111] H.R. 3534, § 208(b)(2).

[112] Data not available for Central Gulf.

In: The National Environmental Policy Act ISBN: 978-1-61324-876-8
Editors: M. Drubich and A. Moser

Chapter 5

THE NATIONAL ENVIRONMENTAL POLICY ACT AND OUTER CONTINENTAL SHELF OIL AND GAS ACTIVITIES*

***National Commission on the BP Deepwater Horizon Oil Spill and Offshore Drilling*†**

This staff working paper focuses on the National Environmental Policy Act of 1969 (NEPA) and its application to offshore oil and gas activities. The paper is divided into three sections. The first section is a general discussion of NEPA as it applies to Federal agencies. The second section addresses the Department of the Interior's (DOI) NEPA regulations and portions of its Departmental Manual that relate to the implementation of those rules for oil and gas activities. The third section focuses on NEPA questions specific to the Deepwater Horizon oil spill.

---

* This is an edited, reformatted and augmented version of a National Commission on the BP Deepwater Horizon Oil Spill and Offshore Drilling publication, Staff Working Paper No. 12.

† Staff working papers are written by the staff of the National Commission on the BP Deepwater Horizon Oil Spill and Offshore Drilling for the use of members of the Commission. They do not necessarily reflect the views of the Commission or any of its members. In addition, they may be based in part on confidential interviews with government and non-government personnel.

I. OPERATION OF NEPA

A. Procedural Nature of NEPA

NEPA establishes a process by which federal agencies are obligated to review the environmental effects of their proposed agency actions during the decision-making process. While NEPA's underlying purpose is to reduce environmental degradation, the nature of the obligation that the statute imposes on agencies to review the impacts of their decisions is essentially procedural. The statute is designed to require federal agencies to prepare formal "statements" in which proposed agency actions are analyzed in a way that promotes informed decision-making rather than a substantive result. Thus, NEPA's goal is not to prevent federal approval of actions that might have environmental consequences, but instead to ensure that agency actions are undertaken knowingly and made in consideration of potential environmental consequences.

The Supreme Court articulated this statutory purpose in the early decades of NEPA's implementation, in the context of its review of NEPA-compelled agency statements. The Court made clear that NEPA is designed to "insure a fully informed and well-considered decision."[1] The statute requires agencies to take a "hard look" at environmental effects of a major federal action,[2] but it does not require agencies "to elevate environmental concerns over other appropriate considerations."[3] NEPA does not bar environmentally-destructive actions,[4] but seeks to promote environmental protection by ensuring rigorous and timely consideration of environmental consequences:

> NEPA does not work by mandating that agencies achieve particular substantive environmental results. Rather, NEPA promotes its sweeping commitment to "prevent or eliminate damage to the environment and biosphere" by focusing Government and public attention on the environmental effects of proposed agency action.[5] By so focusing agency attention, NEPA ensures that the agency will not act on incomplete information, only to regret its decision after it is too late to correct.[6]

The Court also clarified that NEPA is about process and informed decision-making:

> NEPA does not mandate particular results, but simply prescribes the necessary process. . . . If the adverse environmental effects of the proposed action are adequately identified and evaluated, the agency is not

> constrained by NEPA from deciding that other values may outweigh environmental costs. . . . Other statutes may impose substantive environmental obligations on Federal agencies, but NEPA merely prohibits uninformed—rather than unwise—agency action.[7]

B. NEPA's Action-Forcing Requirement — A Detailed Statement

Section 102(2) is the operative provision with respect to NEPA documents.[8] Subsection (A) directs agencies to "utilize a systematic, interdisciplinary approach which will insure the integrated use of the natural and social sciences and the environmental design arts in planning and in decisionmaking which may have an impact on man's environment." The directive for preparation of NEPA documents in association with agency action appears in subsection (C), which requires a "detailed statement by the responsible official." Specifically, subsection (C) requires an agency to consider potential environmental impacts of a proposed action if that action is a "major Federal action[] significantly affecting the quality of the human environment."[9] The agency must adequately discuss not only the environmental impact of the proposed action, but also its unavoidable adverse impacts, alternatives to it, the relationship between short-term uses of the environment and long-term productivity, and any irreversible commitments of resources resulting from the proposed action.[10] In addition, subsection (E) requires the agency to "study, develop, and describe appropriate alternatives to recommended courses of action in any proposal which involve[] unresolved conflicts concerning alternative uses of available resources."[11]

Non-federal action is not subject to NEPA, but can fall under the statute's purview if an agency approves some aspect of the action. Actions that do not have the potential for significant environmental effects are, however, not subject to NEPA's requirement that a federal agency prepare a detailed statement that considers the environmental impacts of the agency's proposed action.[12]

C. CEQ Regulations

Title II of NEPA established the Council on Environmental Quality (CEQ) within the Executive Office of the President. By Executive Order 11991(May

24, 1977), President Jimmy Carter directed CEQ to "[i]ssue regulations to Federal agencies for the implementation of the procedural provisions" of NEPA. The resulting regulations, first promulgated on November 28, 1978, and later amended, set forth the manner by which agencies implement NEPA section 102(2).[13] The regulations, which are binding "except where compliance would be inconsistent with other statutory requirements,"[14] require agencies to establish their own procedures to implement NEPA in consultation with CEQ.[15]

D. NEPA Documents

Under NEPA section 102(2)(C), any agency action that has a potential to significantly affect the quality of the human environment must be accompanied by NEPA review. CEQ regulations employ two documents that outline environmental impacts: an environmental impact statement (EIS) and an environmental assessment (EA).[16] The EIS is the "detailed statement" identified in NEPA to address a major Federal action that has significant impacts on the human environment.[17] By contrast, an EA is a "concise public document" that serves to (1) briefly provide sufficient evidence and analysis for determining whether to prepare an EIS; (2) aid an agency's compliance with NEPA when no EIS is necessary; and (3) facilitate preparation of an EIS when one is necessary.[18]

The point of an EA is to document whether an EIS is required for a proposed action. If an agency knows that an action will generate significant environmental effects, an EIS is required and there is no basis for preparing an EA. But if the agency is unsure, or if the proposed action is of the sort that does not always require an EIS,[19] the agency may prepare an EA. If, as a result of an EA, the agency determines that no significant impact will likely result from the proposed action, the agency issues a "finding of no significant impact" (FONSI).[20] A FONSI may include the discussion underlying its conclusion, or may attach an EA.[21] If an agency issues a FONSI, NEPA compliance is complete because, as a statutory matter, the proposed action will not significantly affect the human environment, and thus a detailed EIS is unnecessary.[22] By contrast, if the analysis in an EA projects a significant impact, an agency cannot issue a FONSI. If an agency wishes to move forward with a proposed action, it must choose an alternative that mitigates impacts such that a FONSI is warranted,[23] or it must prepare a full EIS.[24]

The EIS, EA, and FONSI, are three of four "environmental documents" which effectuate NEPA's mandatory review.[25] The fourth is a Notice of Intent, which is a public notice that an EIS will be prepared.[26] No other NEPA documents exist in the CEQ rules to accomplish the NEPA review required in section 102(2).

E. When and How to Prepare an EIS or an EA

An agency can always choose to prepare an EIS, and must do so if the action is one "normally requiring" an EIS.[27] The scope and format of the EIS is controlled by regulations that require a summary, a section describing the purpose and need for the proposed action, alternatives to it, the affected environment, environmental consequences from the action, and a list of preparers.[28] The EIS generally must be circulated to other agencies and is subject to consultation with relevant members of the public and governmental entities.[29] An EA is less detailed, more concise, and prepared for actions that might have no significant impacts, or ones that can be reduced to a level of "not significant" with mitigation. Considerably less of an investment than an EIS, EAs include "brief discussions of the need for the proposal, of alternatives as required by section 102(2)(E), of the environmental impacts of the proposed action and alternatives, and a listing of agencies and persons consulted." 30

The distinctions between an EA and an EIS give rise to common misconceptions about them. One misconception is that when an agency prepares a full EIS, its work is devoted to preventing all significant impacts. However, by its nature, although an EIS considers significant impacts, it is not a prescription for avoiding them.[31] An agency prepares an EIS *because* significant impacts are anticipated.[32] Therefore, the fact that an EIS identifies adverse impacts is not tantamount to a failure of NEPA compliance. Indeed, in *Robertson v. Methow Valley Citizens Council,* the Supreme Court rejected the idea that an EIS can only be sustained if it provides for detailed mitigation of all potential impacts:

> There is a fundamental distinction, however, between a requirement that mitigation be discussed in sufficient detail to ensure that environmental consequences have been fairly evaluated, on the one hand, and a substantive requirement that a complete mitigation plan be actually formulated and adopted, on the other. . . . [I]t would be inconsistent with NEPA's reliance on procedural mechanisms—as opposed to substantive,

result-based standards—to demand the presence of a fully developed plan that will mitigate environmental harm before an agency can act.[33]

Another misconception is that if an EA is prepared, a FONSI necessarily must result because the EA/FONSI together identify any significant impacts from the proposed action. However, an EA can never constitute final NEPA compliance if it is the vehicle for identifying significant impacts that cannot be mitigated. While it is enough that significant impacts have been fully vetted in an EIS, they must be reduced to a level of "not significant" in an EA for a FONSI to result. An EA is sufficient to conclude NEPA compliance and to lead to a FONSI *only* if it justifies a conclusion that no significant impacts will occur as incremental effects of the proposed action. If this is not the result then further NEPA documentation is required.

The sufficiency of NEPA documentation is determined ultimately by the federal courts in challenges to agency action under the Administrative Procedure Act.[34] A party challenging a NEPA document must demonstrate with objective proof that the agency failed to adequately consider a substantial environmental question of material significance to the proposed action, or otherwise failed to abide by section 102(2) of NEPA. Proof of arbitrary or inconsistent conclusions, or a failure to consider relevant information, demonstrates a lack of compliance with NEPA because it proves an "arbitrary and capricious" action under the Administrative Procedure Act.[35]

An EIS is judged by whether it constitutes a "detailed statement" that takes a "hard look" at all of the potential significant environmental consequences of the proposed action and reasonable alternatives thereto, considering all relevant matters of environmental concern.[36] The question is whether the EIS contains a "reasonably thorough discussion of the significant aspects of the probable environmental consequences" of the proposed action and alternatives.[37] Significant impacts do not defeat an EIS under the Administrative Procedure Act.

By contrast, if an EA projects significant effects, a FONSI cannot be sustained and a challenge under the Administrative Procedure Act[38] will prevail. An EA for which a FONSI has been issued will be affirmed when the record demonstrates that the agency has considered relevant matters of environmental concern, taken a "hard look" at potential environmental impacts, and made a convincing case that no significant impact will result or that any impact will be reduced to insignificance by the adoption of appropriate mitigation measures.[39]

F. Categorical Exclusions—Agency Actions Exempt from the Action-Forcing Requirements of NEPA

Preparation of an EA or EIS that analyzes the effects of a proposed action and its reasonable alternatives is a procedural line an agency must cross before it can move forward with an action. CEQ has only included the EIS, EA, resulting FONSI (EA), and Notice of Intent (of an EIS) on its list of NEPA documents.[40] CEQ understood, however, that NEPA does not, and should not, compel unnecessary paperwork.[41] Therefore, CEQ identified "categorical exclusions" (CE)—sets of agency actions that do not generate environmental effects and thus are exempt from the case by case application of section 102(2). CEQ allows each agency to establish exceptions from the NEPA process for entire categories of actions, "which do not individually or cumulatively have a significant effect on the human environment and which are therefore exempt."[42] CEQ defines a CE as that "category of actions which do not . . . have a significant effect on the human environment and which have been found to have no such effect in procedures adopted by a Federal agency in implementation of the regulations[43]and for which, therefore, neither an [EA] nor an [EIS] is required. . . . Any procedures under this section shall provide for extraordinary circumstances in which a normally excluded action may have a significant environmental effect."[44] A CE is thus *not* a level of NEPA review, but rather a defined category of agency action that does not require NEPA review apart from the development of the CE itself..

CEs represent the common sense notion that agencies routinely "act" in ways that do not create environmental effects requiring a "hard look." For example, employee performance evaluations or organizational changes do not generate impacts on the physical world and thus are not subject to NEPA. In *Sabine River Authority v. Department of the Interior*,[45] the Fifth Circuit held that "the acquisition of a negative easement which by its terms prohibits any change in the status quo does not amount to 'major Federal action[] significantly affecting the quality of the human environment.'" Although, in this case, the court did not specifically define an agency-asserted CE, it did define a subset of actions exempt from NEPA as actions that do not change the physical world. The fact that an agency must examine a proposed action to determine whether it may properly be within a defined CE is an activity occasioned by NEPA, but the examination itself is not NEPA review in a NEPA document as defined by CEQ.

Agencies must establish a list of their own CEs, which are defined to be "categories of actions which do not individually or cumulatively have a significant impact on the human environment and are therefore exempt" from NEPA.[46] Environmental documentation in the form of an EIS or EA is required unless an agency has established a procedure that includes identification of "typical classes of action . . . [w]hich normally do not require either an [EIS] or an [EA]."[47]

G. Effects/Impacts

In addressing the environmental consequences of a proposed action in an EIS, an agency must discuss both its direct and indirect effects as well as alternatives to the proposed action in order to offer meaningful comparisons to other actions that may also be implemented.[48] Among the effects to be addressed are cumulative impacts[49] and ways to mitigate effects.[50] Similarly, EAs must include brief statements of "environmental impacts" of a proposed action.[51]

"Effects and impacts as used in these regulations are synonymous" and include "direct," "indirect," and "cumulative" effects.[52] Direct effects are "caused by the action and occur at the same time and place."[53] Indirect effects are "caused by the action and are later in time or further removed in distance, but are still reasonably foreseeable. Indirect effects may include growth inducing effects and other effects related to induced changes in the pattern of land use, population density or growth rate, and related effects on air and water and other natural systems, including ecosystems."[54] A "reasonably close causal relationship" between the Federal action and the effects at issue is critical; where the chain of causation is unduly remote, an effect will not be deemed to be an indirect effect of a proposed action.[55] A cumulative impact is the impact on the environment that results from the incremental impact of the action when added to other past, present, and reasonably foreseeable future actions, regardless of what agency or person undertakes such other actions.[56] Cumulative impacts can result from individually minor but collectively significant actions taking place over time.[57]

When certain impacts are to be analyzed depends on whether they may be reasonably foreseeable.

H. Worst Case Analysis

There is general confusion about whether agencies are required to consider a "worst case scenario" in their NEPA documents. At one time, regulations did require EISs to include worst case scenario analyses in certain circumstances. But CEQ eliminated this requirement in 1986 by revising its NEPA regulations. Worst case scenarios are not currently required in the NEPA regulations.

The worst case analysis requirement first arose in the 1978 CEQ regulations. CEQ was attempting to resolve how an agency should handle situations in which the impacts of an action were uncertain or unknown. It instructed federal agencies to obtain additional data when addressing "scientific uncertainty" or "gaps in relevant information" regarding significant adverse environmental effects that were "essential to a reasoned choice among alternatives" or "important to the decision." [58] The problem was that some such data was usually not available at all, or if available, unreasonably costly to obtain. Accordingly, if additional and necessary scientific information could not be acquired because the "overall costs of obtaining it were exorbitant," or "the means for obtaining it were beyond the state of the art," the agency was required to "include a worst case analysis and an indication of the probability or improbability of its occurrence."[59]

CEQ amended the regulation in 1986. The change was controversial and generated considerable public comment. CEQ responded by explaining that it believed the 1978 regulation was "flawed" and that it needed to provide "better and more logical" requirements for agencies to deal with situations involving incomplete information.[60] CEQ disagreed with the view expressed by certain commenters that removing the worst case scenario requirement would weaken NEPA compliance, instead stating that a "less sensational" and "more careful and professional approach" was a "better" way of dealing with incomplete and unavailable information in an EIS.[61]

In *Robertson v. Methow Valley Citizens Council,* the Supreme Court reversed a circuit court that had insisted the 1978 regulation's demand for a worst case analysis be followed even after amendment of that regulation. The Supreme Court gave "substantial deference" to the 1986 regulatory amendment, citing the fact that the 1978 regulation had been subject to considerable criticism. Citing the 1985 proposed rule that became the 1986 amendments, the Court stated:

> CEQ explained that by requiring that an EIS focus on reasonably foreseeable impacts, the new regulation "will generate information and discussion on those consequences of greatest concern to the public and of greatest relevance to the agency's decision," 50 Fed. Reg. 32237 (1985), rather than distorting the decision-making process by overemphasizing highly speculative harms, 51 Fed. Reg. 15624-15625 (1986); 50 Fed. Reg. 32236 (1985). In light of this well-considered basis for the change, the new regulation is entitled to substantial deference. Accordingly, the Court of Appeals erred in concluding that the [relevant EIS] is inadequate because it failed to include a "worst case analysis."[62]

The regulation in place today states that if "information relevant to reasonably foreseeable significant adverse impacts cannot be obtained" because it is too costly to get or unavailable, the agency must include a four-part statement regarding the information, including "the agency's evaluation of such impacts based on theoretical approaches or research methods generally accepted in the scientific community."[63] "Reasonably foreseeable" environmental effects include "impacts which have catastrophic consequences, even if their probability of occurrence is low, provided that the analysis of the impacts is supported by credible scientific evidence, is not based on pure conjecture, and is within the rule of reason."[64]

I. Consultation

CEQ regulations establish extensive requirements for including other agencies and the public in the EIS process. CEQ's stated policy is to encourage and facilitate "public involvement in decisions which affect the quality of the human environment."[65] The rules "emphasize agency cooperation early in the NEPA process."[66] In addition, they require "diligent efforts to involve the public in preparing and implementing their NEPA procedures."[67]

The requirements for public and agency review of a pending EIS are demanding. The lead agency responsible for the proposed action may request "any other Federal agency which has jurisdiction by law," or any such agency "which has special expertise with respect to any environmental issue" to be addressed in the EIS, to be a "cooperating agency."[68] The cooperating agency's involvement and participation should be solicited at the earliest possible time.[69] The regulations call for "scoping," which is the "early and open process for determining the scope of issues to be addressed and for identifying the significant issues related to a proposed action."[70] As soon as it decides to prepare an EIS, the agency must publish a "notice of intent" in the Federal Register, subject to limited exceptions.[71] The notice of intent invites "participation of affected Federal, State, and local agencies, any affected Indian tribe, the proponent of the action, and other interested persons."[72]

After the scoping process, the EIS is issued in two stages. A draft environmental impact statement (DEIS) will issue "in accordance with the scope decided upon in the scoping process."[73] The DEIS should discuss "all major points of view on the environmental impacts of the alternatives including the proposed action."[74] At this juncture, the agency must invite comments.[75] With respect to "any Federal agency that has jurisdiction by law or special expertise with respect to any environmental impact involved, or which is authorized to develop and enforce environmental standards," the lead agency shall obtain comments."[76] It "shall request" comments of appropriate state and local agencies authorized to develop and enforce environmental standards; Indian tribes with a reservation that may suffer impacts; any agency which has requested statements on actions of the kind proposed; any applicant; and the public.[77] The agency must "affirmatively solicit[] comments from those persons or organizations who may be interested or affected."[78]

The agency then prepares the final environmental impact statement (FEIS). In doing so, the agency must "assess and consider comments both individually and collectively, and shall respond in one or more" of the following ways: (1) modify alternatives including the proposed action; (2) develop and evaluate alternatives not previously considered; (3) supplement, improve, or modify analyses; (4) make factual corrections; or (5) give a reasoned explanation for why the comments do not warrant further response.[79]

Several directives are clear from CEQ regulations regarding consultation with respect to an EIS. The agency is required to include federal, state, local, and tribal agencies, and the public, as well as to respond to commenters.[80] The agency is not required to agree with the commenter, so long as the agency discloses an opposing point of view and reasonably responds to it. On the one hand, "disagreement among experts will not serve to invalidate an EIS."[81] On the other, a failure to disclose and respond to a commenter's evidence and opinions directly challenging the scientific basis upon which the EIS rests will defeat it.[82] The Ninth Circuit has explained that the NEPA requirement to address opposing viewpoints in an FEIS "reflects the paramount Congressional desire to internalize opposing viewpoints into the decision-making process to ensure that an agency is cognizant of all the environmental trade-offs that are implicit in a decision."[83] Once again, the NEPA burden is procedural: the agency must disclose and respond. The Administrative Procedure Act requires that the response must make sense; otherwise, the agency would reasonably be accused of acting in an arbitrary and capricious manner. Accordingly, a federal agency's comments may be given extra weight over and above that of an ordinary citizen, depending on the agency's expertise in the topic.

The obligations for public disclosure and comment with respect to an EA are entirely different. Unlike the rigid public/governmental comment requirements with respect to an EIS, agencies need only involve "environmental agencies, applicants, and the public" in preparing EAs "to the extent practicable."[84] Normally, the obligation of the agency in preparing a FONSI is to disseminate it to the affected public.[85] In "certain limited circumstances," however, the agency must make the FONSI "available for public review (including [through] State and areawide clearinghouses) for 30 days before the agency makes its final determination whether to prepare an [EIS] and before the action may begin."[86] There are two "limited circumstances": where the proposed action "is, or is closely similar to, one which normally requires the preparation" of an EIS, or where its nature is "without precedent."[87]

The federal courts have been inconsistent in their interpretations of the CEQ regulations with respect to public involvement in the EA process. On more than one occasion, the Ninth Circuit has held that a failure to involve the public constitutes a violation of the CEQ regulations. In *Citizens for Better Forestry v. U.S. Department of Agriculture,* the court stated: "Although we have not established a minimum level of public comment and participation required by the regulations governing the EA and FONSI process, we clearly have held that the regulations at issue must mean something. . . . It is evident, therefore, that a complete failure to involve or even inform the public about an agency's preparation of an EA and a FONSI, as was the case here, violates these regulations." [88] And in *Anderson v. Evans*, the Ninth Circuit held that "[t]he public must be given an opportunity to comment on draft EAs and EISs." [89]

By contrast, the First Circuit has held that CEQ's public participation requirements are satisfied even when an agency has not offered a draft or final EA for public comment on grounds that "[n]othing in the CEQ regulations requires circulation of a draft EA for public comment, except under certain 'limited circumstances.'"[90] Similarly, the Tenth Circuit upheld an agency's publication of an EA when the agency issued public notice of the proposed action and provided a comment period, even though the record did not clarify whether the EA was made available to the public before the agency issued its decision.[91] And in *Pogliani v. U.S. Army Corps of Engineers*,[92] the Second Circuit found that the plaintiffs were unlikely to succeed on their claim that the Corps erred by failing to release a draft EA and FONSI for public comment prior to its issuance in final form.

The District of Columbia federal district courts have inconsistently interpreted CEQ regulations. In *Fund for Animals v. Norton,*[93] the U.S. District Court for the District of Columbia struck down as inadequate the publication by the U.S. Fish and Wildlife Service of the notice of the availability of a draft EA in the Federal Register, which afforded the public a two-week comment period followed two weeks later by finalization of the EA, and a week later by issuance of the FONSI. By contrast, the same court concluded that "an agency is *not* expressly required to circulate a draft EA for public comment before adopting its final decision, except in limited circumstances that do not apply here."[94] The court rejected the plaintiff's argument that the agency was required to provide public opportunity to comment on a draft EA, noting that the CEQ regulation only required publication "to the extent practicable."[95] Accordingly, the court focused on the factual inquiry as to what had been "practicable" for the agency to accomplish in the EA process.

J. Timing—Supplemental Statements and Tiering

Generally, the appropriate time for preparing a NEPA document is prior to a decision, when the decision-maker has the widest range of options.[96] When the decision results in an "irreversible and irretrievable commitment of resources," NEPA review is required.[97] One of the most difficult NEPA compliance questions presented to an agency is how to approach agency action that is ongoing, repetitive, or takes place in stages. Arguably, the largest federal programs with impacts on the physical world are always ongoing. The CEQ regulations address the process of considering impacts in consecutive NEPA document as "supplementing" NEPA review or "tiering" one NEPA document to another. In such cases, an earlier NEPA document is incorporated into a latter one.

Agencies must prepare "supplements to either draft or final [EISs]" if the agency "makes changes in the proposed action that are relevant to environmental concerns" or if "[t]here are significant new circumstances or information relevant to environmental concerns and bearing on the proposed action or its impacts."[98] The supplementing process is further described as follows:

> [A]n agency need not supplement an EIS every time new information comes to light after the EIS is finalized. To require otherwise would render agency decision-making intractable, always awaiting updated information only to find the new information outdated by the time a decision is made. On the other hand, and as the petitioners concede, NEPA does require that agencies take a "hard look" at the environmental effects of their planned action, even after a proposal has received initial approval. . . . Application of the "rule of reason" thus turns on the value of the new information to the still pending decisionmaking process. In this respect the decision whether to prepare a supplemental EIS is similar to the decision whether to prepare an EIS in the first instance: If there remains "major Federal actio[n]" to occur, and if the new information is sufficient to show that the remaining action will "affec[t] the quality of the human environment" in a significant manner or to a significant extent not already considered, a supplemental EIS must be prepared.[99]

If information uncovered after a NEPA statement is sufficiently new and significant when compared to the information upon which the NEPA statement was based, a new NEPA statement is required.

The related concept of "tiering" addresses the sequencing of environmental statements generally from broad documents to narrower statements or analyses, or from a broad programmatic to a site-specific analysis. Tiering is defined in the CEQ regulations as "coverage of general matters in broader [EISs] . . . with subsequent narrower statements or environmental analyses . . . incorporating by reference the general discussions and concentrating solely on the issues specific to the statement subsequently prepared."[100] The point of tiering is to allow an agency to avoid duplicating paper and effort. Assuming a broader NEPA document considers impacts, a narrower NEPA document can incorporate such consideration by reference. Impacts must be considered in some document, however, and if the agency does tier, the question is whether the analysis in the first NEPA document is sufficient to support tiering in the second one. In *Kern v. Bureau of Land Management*,[101] the Ninth Circuit held that an environmental impact, consideration of which is eschewed in an EA, must necessarily have been addressed adequately in a NEPA document to which the EA is tiered.

As long as the necessary analysis of a particular action is undertaken at a relevant action point, supplementation and tiering are entirely accepted and common practices. In either case, a NEPA statement that is tiered to a final

EIS, or which is supplemental to one, need not restate impacts analysis or alternatives that were already considered in the document it supplements or to which it is tiered. Nonetheless, a NEPA document may only be tiered under the CEQ regulations to a document that has itself been issued as a document under NEPA.[102] Likewise, supplementation of NEPA documents can only occur in NEPA documents.

Because broad Federal programs are ongoing, tiering and supplementation can often become the rule rather than the exception. Questions surrounding such programs include when to begin a new programmatic review, when a new NEPA review may be required, and what requirements are imposed when supplemental or new information arises. In simplistic terms, if significant new information comes to light, a supplemental review document is required that could be tiered to a prior document. The staging of NEPA review cannot ultimately be employed to evade NEPA review of a particular impact or action. As sensible as these concepts may sound, the staging of NEPA review, and the avoidance of review that staging may unwittingly permit, are difficult questions for agencies to resolve.

K. Agency-Formulated Documents that Result from NEPA Compliance But Do Not Constitute NEPA Review

NEPA compliance in the context of ongoing agency actions has given rise to certain practical problems for agencies attempting to determine whether they need to supplement NEPA analysis when NEPA documents already exist. If an agency undertakes actions that have already been subject to NEPA review and therefore do not require further NEPA analysis, the agency may nonetheless be challenged for moving forward without preparing any NEPA document. For this reason, agencies have established procedures outside the scope of the CEQ regulations to determine whether NEPA analysis already undertaken is sufficient such that no further review is required. Federal courts have upheld such agency procedures. In *Pennaco Energy v. U.S. Department of the Interior*,[103] the Tenth Circuit addressed such a document—in this case, a Documentation of NEPA Adequacy—that had been prepared by DOI's Bureau of Land Management (BLM) to determine what NEPA analysis existed, and whether more NEPA review should be undertaken to supplement it. While noting that "[Documentation of NEPA Adequacy], unlike EAs and FONSIs, are not mentioned in the NEPA or in the regulations implementing the NEPA," the court upheld BLM's use of "non-NEPA procedures to determine whether new NEPA documentation is required."[104] Though such documents cannot supplement previous EAs or EISs or address site-specific effects not previously considered in them, they identify for the decision-maker the location of existing NEPA analysis, and allow the decision-maker to determine whether more is needed.

Other agencies employ different terms for procedures to assess NEPA analysis.[105] The critical point to note about these approaches is that, although they are legal means to assess the sufficiency of existing NEPA review; these procedures and documents are not NEPA documents and do not constitute NEPA review. Whether NEPA review is sufficient, preparation of a Documentation of NEPA Adequacy-type document depends on whether NEPA supplementation is required.

II. DOI'S IMPLEMENTATION OF NEPA PROCEDURES

Consistent with CEQ's directives, DOI has promulgated regulations and guidance to ensure proper application of NEPA and CEQ's binding rules. The regulations appear at 43 C.F.R. § 46, and generally reproduce and supplement CEQ's directions. In addition, DOI has prepared a Departmental Manual that establishes guidance, direction, and requirements for, *inter alia*, the proper application of NEPA requirements to Departmental action. In addition to general Departmental guidance, the Departmental Manual gives specific direction to each bureau within the Department, including the former Minerals Management Service (MMS), which has been succeeded since the Macondo well blowout last spring by the Bureau of Ocean Energy Management and Regulatory Enforcement (BOEMRE).[106] The regulations and Manual provisions are detailed and lengthy. To avoid repeating the above analysis of NEPA, the following section mentions only provisions of note or interest to the Commission's work.

A. DOI NEPA Rules

The regulations at 43 C.F.R. § 46 ("Implementation of [NEPA]") generally repeat the requirements outlined in the CEQ rules. The Departmental rules make clear that use of preexisting NEPA documentation, when further NEPA review would merely duplicate prior statements, should be a priority.[107] The regulations address "incorporation of NEPA documents" and "tiered documents."[108] They also identify a document not mentioned in the CEQ regulations—a Finding of No New Significant Impacts (FONNSI). This document must follow the preparation of an EA and is distinguished from a FONSI in that it signifies the EA is tiered to a prior NEPA document (specifically an EIS), and the EA process has revealed that all significant environmental impacts were studied and any new impacts identified are not significant. The full text of this regulation is as follows:

> An environmental assessment prepared in support of an individual proposed action can be tiered to a programmatic or other broader-scope environmental impact statement. An environmental assessment may be prepared, and a finding of no significant impact reached, for a proposed action with significant effects, whether direct, indirect, or cumulative, if the environmental assessment is tiered to a broader environmental impact statement which fully analyzed those significant effects. Tiering to the programmatic or broader-scope environmental impact statement would allow the preparation of an environmental assessment and a finding of no significant impact for the individual proposed action, so long as any previously unanalyzed effects are not significant. A finding of no significant impact other than those already disclosed and analyzed in the environmental impact statement to which the environmental assessment is tiered may also be called a "finding of no new significant impact."[109]

In its entirety, the regulation is accurate, but the italicized portion could be read as inconsistent with the CEQ regulations if taken out of context. It is not accurate to say that a FONSI can be reached for a proposed action with significant effects, even though it is true that a FONSI can be reached if all effects from the proposed action are addressed in a document to which the EA is tiered. By contrast, if incremental effects specifically caused only by the proposed action addressed in the EA are significant, and are not directly addressed in the EIS, a FONSI cannot be issued. The regulation's creation of a new type of document (FONNSI) may confuse the proper application of NEPA in the case of tiering or supplementation for the public and for agency employees obligated to implement NEPA. Whether an effect of the action studied in the second NEPA document is fully addressed in the first NEPA document may be subject to debate.

The DOI's rules also address "[c]onsultation, coordination, and cooperation with the agencies."[110] They appear to expand the consultation requirement established in the CEQ rules by directing consultation "whenever possible" for all NEPA documentation, not just EISs:

> The Responsible Official must whenever possible consult, coordinate, and cooperate with relevant State, local, and tribal governments and other bureaus and Federal agencies concerning the environmental effects of any Federal action within the jurisdictions or related to the interests of these entities.[111]

The DOI NEPA rules provide specific requirements for consultation and notification in association with a Notice of Intent, the precursor to the EIS.[112] Nonetheless, regulations specifically addressing EAs avoid any suggestion that these rules expand the CEQ consultation requirements for EAs. Instead, 43 C.F.R. § 46.305 repeats CEQ's requirement that consultation is necessary only "to the extent practicable"; specifies that doing so is at the discretion of the Responsible Official[113]; ensures that the Department is not required to issue draft EAs and may do so only if the Official believes it to be "appropriate"; and permits the decision-maker to consult with Cooperating Agencies. The only mandatory requirements in the rules are to consider comments given, even if not solicited, and to "notify" the public of the final EA and FONSI. Comments on a FONSI generally do not need to be solicited, except as required by 40 C.F.R. § 1501.4(e)(2), a CEQ regulation that requires a 30-day notice for a FONSI to be published in "limited circumstances" defined by individual agencies.[114]

DOI's rules detail the preparation of EAs and EISs in ways more extensive than the CEQ rules. Subpart D addresses EAs.[115] Subpart E addresses EISs.[116]

The regulations also address CEs. Like the CEQ regulations, the Department defines a CE as "a category or kind of action that has no significant individual or cumulative effect on the quality of the human environment."[117] The rule goes on to note that "[e]xcept as provided in paragraph (c) of this section, if an action is covered by a Departmental CE, the bureau is not required to prepare" an EA or EIS.[118] The reference to paragraph (c) is to the "extraordinary circumstances under which actions otherwise covered by a categorical exclusion require analyses under NEPA," which are identified at 43 C.F.R. § 46.215. "Any action that is normally categorically excluded must be evaluated to determine whether it meets any of the extraordinary circumstances in section 46.215; if it does, further analysis and environmental documents must be prepared for the action."[119] Section 46.210 lists actions that constitute "Department-wide" CEs. Though the list is extensive, and includes personnel decisions and organizational changes, no items on the list pertain to management of oil and gas resources by any bureau.

The Departmental regulations establish the extraordinary circumstances that make an action otherwise a potential CE subject to NEPA review. "Extraordinary circumstances" exist for actions that:

a) Have significant impacts on public health or safety.
b) Have significant impacts on such natural resources and unique geographic characteristics as historic or cultural resources; park, recreation or refuge lands; . . . wetlands (EO 11990); floodplains (EO 11988); . . . migratory birds; and other ecologically significant or critical areas.
c) Have highly controversial environmental effects . . .
d) Have highly uncertain and potentially significant environmental effects or involve unique or unknown environmental risks.
e) Establish a precedent for future action or represent a decision in principle about future actions with potentially significant environmental effects.
f) Have a direct relationship to other actions with individually insignificant but cumulatively significant environmental effects.
g) Have significant impacts on species listed, or proposed to be listed, on the List of Endangered or Threatened Species or have significant impacts on designated Critical Habitat for these species.[120]

Notably, the Department's regulations add a category of exclusions from NEPA review not found in the CEQ rules—actions excluded by statute. "Congress may establish categorical exclusions by legislation, in which case the terms of the legislation determine how to apply those categorical exclusions."[121]

B. Departmental Manual

For decades, DOI has issued "instructions, policies, and procedures that have general and continuing applicability to Departmental activities, or that are important to the management of the Department" in a Departmental Manual.[122] The Departmental Manual describes the organizations and functions of the Department's bureaus and offices, delegations of the Secretary's authority, and the policies and general procedures for administrative activities and the operations of specific programs. It is used to communicate the instructions of the Office of the Secretary throughout the

Department, to provide guidance to the bureaus and offices in their administrative and program operations, and to serve as the primary source of information on organization structure, authority to function, and policy as well as general procedures. Bureaus and offices must comply with the provisions of the Departmental Manual, except to the extent that the provisions are superseded by appropriate authority (e.g., a change in statute, regulation, an executive order; a secretary's order, or a court decision).[123]

i. NEPA Procedures at Part 516 of the Departmental Manual

The Department first proposed draft procedures to implement NEPA and the CEQ regulations on July 10, 1979.[124] The first procedures were adopted as Part 516 of the Departmental Manual on March 18, 1980, to provide "policy and procedural directions to all elements of the Department for complying with the mandate of NEPA and the CEQ regulations."[125] The Department adopted a broad list of "Departmental Categorical Exclusions" in the Departmental Manual, but explained that "[i]nstructions specific to bureaus will be issued [subsequently] as appendices to these procedures." 126 Adopting the language from the procedures proposed in 1979, the Department established criteria "to determine actions to be categorically excluded from the NEPA process."[127] The criteria are as follows:

(a) The action or group of actions would have no significant effect on the quality of the human environment.
(b) The action or group of actions would not involve unresolved conflicts concerning alternative uses of available resources.[128]

The Departmental Manual goes on to note:

> Based on the above criteria, the classes of actions listed in Appendix 1 to this Chapter are categorically excluded Department-wide from the NEPA process, Bureaus shall list categorical exclusions specific to their programs in the Bureau Appendix to Chapter 6.[129]

In Chapter 6, the Departmental Manual directs "heads of Bureaus and offices to "propose categorical exclusions."[130] As described below, these provisions remained in place until 2009, when Chapters 1-6 of the Departmental Manual were amended and all references to CEs and exceptions to their use were removed.

ii. Departmental Manual Offshore CEs

From the first Departmental oil and gas NEPA compliance review, the United States Geological Survey (USGS) defined certain key approvals in the Western and Central Gulf of Mexico as categorically excluded from NEPA review, including: applications for permits to drill (APD); exploration plans (EP); and development and production plans (DPP); and then, as amended later, development operations and coordination documents defined by MMS.[131]

The USGS's decision to treat Gulf of Mexico planning documents and the drilling of wells in this region as exempt from NEPA review appears to have stemmed from statutory provisions in, and congressional commentary associated with, passage of the Outer Continental Shelf Lands Act (OCSLA) amendments of 1978. Although USGS has never explained its decision to treat EPs as categorically excluded, USGS may have believed that DPPs were statutory CEs, which could not be subject to regulatory exception. USGS also treated APDs in the Western Gulf region the same as it treated other applications for the Outer Continental Shelf—categorically excluded as effectively "tiered" to approved planning documents. This conclusion was a questionable application of the CE, as defined by CEQ, for the entire Outer Continental Shelf, but improper in the Western Gulf because no tiering to a CE (a non-NEPA document) is allowed. MMS later came to believe that planning documents in the Western Gulf were categorically excluded from NEPA review because they met the CEQ definition of a CE,

even though MMS regulations acknowledged that such planning document approvals *did* have significant impacts on the human environment in other Outer Continental Shelf regions. MMS squared neither its understanding of how the USGS CEs came into being nor its contradictory description of the effects on the human environment of approvals of the same planning documents, as occurring or not occurring depending on a geographic boundary.

By the time the relevant approvals were given for the Macondo well, all of these problems undercut the use of CEs. Even if the categorical exceptions, as applied to APDs and planning documents in the Western Gulf, could be squared today, it is questionable whether MMS properly applied these exceptions to the circumstances presented by deepwater wells.

The convoluted history of defined CEs for offshore oil and gas development stems from a statute, is developed through the Departmental Manual, and coincides with regulations enacted by relevant bureaus. The basis for the Department's determination that the APDs, EPs, DPPs, and DOCDs in the Western Gulf are categorically excluded from NEPA review seems to have been poorly understood by the MMS. In particular, Notices to Lessees changed the application of the rules in critical ways.[132] Provisions of regulations cited in the Departmental Manual have been superseded, compounding problems with interpreting what does and does not apply offshore. The following discussion attempts to extract the story of the creation and maintenance of the CEs by examining the text and legislative history of the OSCLA 1978 amendments, the pertinent parts of the Departmental Manual, and the Departmental regulations. The discussion reveals seeming inconsistencies with the discussion of NEPA in Part 1 of this paper. Accordingly, the following points must be kept in mind:

1) Other than citing to the 1978 amendments to OCSLA and regulations implementing that statute, how the Department derived CEs for DPPs (or DOCDs) in the Western Gulf region was not articulated when they were created.
2) Staff research has not determined why the USGS concluded that approval of EPs for the Western Gulf of Mexico region should constitute a CE.
3) To the extent the Department defined Western Gulf EPs and DPPs as CEs based on its reading of the OCSLA amendments of 1978, this was not a result of a statutory mandate by Congress.
4) If MMS believed that relevant CEs were in place because of a statutory mandate, it would be inappropriate for it to undertake to review whether exception to the CE should occur because of extraordinary circumstances, because the exception could not override a statutory provision.
5) Approval of the planning documents defined to be CEs has not been specifically identified by any Departmental bureau not causing impacts on the environment, and therefore fitting within the CEQ definition of a CE.
6) If the Department believed the relevant CEs were in place because they met the CE definition in the CEQ rules or Departmental regulations, then any approval (APD, EP, DPP, DOCD) would have to be reviewed to ensure that a particular action is not an unusual circumstance calling for NEPA review before being categorized as a CE.
7) APDs are categorically excluded from NEPA review if the subject wells are identified in an approved plan (EP, DPP, DOCD) that,

outside the Western Gulf region, would be subject to NEPA review. In the Western Gulf region, the plans to which an APD is "tiered" are not subject to NEPA review because they are identified as CEs. Such "tiering" to non-NEPA documents is improper.

C. The Source of the Western Gulf of Mexico Offshore Categorical Exclusions— OCSLA Amendments of 1978

In 1978, Congress enacted Public Law 95-372 to amend OCSLA. A single provision, section 25, is the source of the CEs in the Departmental Manual. Section 25 (codified at 43 U.S.C. § 1351), which addresses "oil and gas development and production," is particularly relevant for two mains reasons. First, it adds a requirement that OCS lessees must submit a detailed DPP for approval by DOI, which, at least once in an area or region, shall be considered a "major Federal action significantly affecting the quality of the human environment" and thus require an EIS.[133] Section 25 establishes extensive consultation requirements for these plans and terms for coordination with coastal zone management provisions. Second, section 25 explicitly excepts the Gulf of Mexico region from its terms. In an exception to *that* exception, Congress added a final subsection (l) that permits the Secretary of the Interior to apply section 25 to leases in the Eastern Gulf of Mexico.[134] The end result is that the OCSLA, as written today, requires Outer Continental Shelf lessees to submit DPPs, and requires that at least once for each area such a plan must be subject to an EIS, except in the Western Gulf of Mexico, where these requirements do not apply.

The legislative history surrounding the OCLSA 1978 amendments shows that competing factors produced these amendments. High energy prices in the late 1970s combined with America's increasing dependency on foreign oil prompted a call for increased development in undeveloped Outer Continental Shelf areas, and for a quick turnaround on actual production once the leases were issued. As the legislative history explained:

> Until December 1975, the OCS leasing program was confined almost exclusively to the Gulf of Mexico. The only exception was a small area leased off southern California in the Santa Barbara Channel. [...] If we are to increase our OCS oil and gas development, leasing must take place in new or "frontier" areas.[135]

The history of the amendments further describes "steps already taken in that direction," including President Carter's directive that CEQ study environmental impacts of oil and gas production on the Atlantic and Gulf of Alaska Outer Continental Shelf, an acceleration in leasing from 1 million acres in 1973 to 10 million acres in 1975, and a DOI Federal Register request for comment on 17 potential Outer Continental Shelf oil and gas leasing areas.[136] Concern also stemmed from the 1969 Santa Barbara oil spill. Congress wanted to ensure increased oversight of development and production, through the plans identified in section 25, with concomitant environmental protection, state and local consultation provisions, and coordination with Coastal Zone Management Act requirements. This was "one of the most important provisions."[137] Each lessee was required to submit a DPP with "information about the nature and extent of the proposed development, including . . . environmental safeguards and safety standards to be implemented."[138] The DPP was subject to consultation with the governors, local officials, and the public.[139]

The industry expressed concerns about additional environmental review, stating that it was a "constraint[] to expanded production."[140] The Government Accounting Office reported that industry representatives demanded "timely, efficient, and effective methods for environmental assessment and realistic assessment of tradeoff[s] between energy needs and environmental hazards."[141] Indeed, the American Petroleum Institute suggested that the additional NEPA and consultation requirements could add three to six years to the time between leasing and production.[142]

Section 25 manifested something of a compromise. In the OCSLA 1978 amendments, DPPs subject to NEPA and consultation requirements were compulsory in frontier areas, but not in the Gulf of Mexico or the Santa Barbara Channel, where development was well-established.[143] The legislative history includes correspondence illustrating a debate about these issues among Congress, Federal agencies, the Secretary of the Interior, and the industry. In a letter to the Senate Subcommittee on Energy and Natural Resources Chief Counsel dated May 10, 1977, the Secretary of the Interior explained the reasoning behind the exception for lessees in the Gulf and Santa Barbara Channel, but expressed concerns about this exception. He proposed alternative language that would permit DOI to make case-by-case determinations as to whether a DPP might be advisable even in the excepted regions:

> One of the most significant changes in S. 9 and H.R. 1614 of existing practice is the detailed review and state participation required prior to

> approval of development and productions plans. Approval of these plans will in many cases require preparation of an [EIS]. An EIS would be required, for example, at least once prior to major development in any area or region of the OCS. These are addressed in section 25 of the bills. Section 25, however, excludes the Santa Barbara Channel and the Gulf of Mexico from its requirements. We believe the Secretary should have the option, where necessary, of applying the provisions of section 25 in these two areas. The Santa Barbara Channel and the Gulf of Mexico have been well developed within recent years and the purpose for the general exclusion in section 25 is understandable. There may be, however, more areas of the Channel or the Gulf where application of these provisions is advisable. Therefore, we recommend addition of the following language as section 25(k) . . . .
>
> "An oil and gas lease issued or maintained under this Act which is located in the Gulf of Mexico or the Santa Barbara Channel shall be subject to the provisions of this section if the Secretary determines, pursuant to regulations prescribed by [him], that the likely environmental or onshore impacts of the development and production of such lease make the application of the provisions of this section in the public interest."[144]

The Secretary of the Interior's suggested provision was not adopted. In the final version of the OCSLA amendments of 1978, the Santa Barbara exception was removed, and the final subsection (l), ensuring that Department could apply section 25 in the Eastern Gulf of Mexico, was added. The Western Gulf, however, remained an exception to any requirement of a DPP, which in section 25 was "at least once" the subject of an EIS. As the August 10, 1978, Conference Report explained:

> The House amendment requires a development and production plan to be submitted for all future leases in a frontier area. The Senate bill provides for a development and production plan to be submitted for all future leases anywhere. The conference report requires a plan to be submitted for all future leases except in the Gulf of Mexico.
>
> The House amendment also requires a plan to be submitted for existing leases in frontier areas, where no oil or gas has been yet discovered. The Senate bill similarly requires a *plan to be submitted for existing leases where there has not yet been a discovery, but exempts the Gulf of Mexico. The conference report adopts the Senate language. Thus, the mandate and specific procedures of this bill that the Secretary of Interior must secure submission, and then review, approve, or disapprove a development and production plan applies to new leases or existing*

> *leases where there has not yet been a discovery and does not apply to leases, old or new, in the Gulf of Mexico.*
>
> *The requirements of this new section are specifically made inapplicable to the Gulf of Mexico. However, there are areas in the eastern gulf that have never been developed. . . . The conferees therefore adopted a provision that gives the Secretary of the Interior the* discretion to require submission of plan[s]–in accordance with this section–for development and production activities in this area which is defined as being adjacent to the State of Florida.[145]

The statute, as amended by section 25 of the OCSLA amendments of 1978, remains in effect. Subsection (b) prohibits development of a lease anywhere on the OCS "other than the Gulf of Mexico" without a DPP.[146] Subsection (e)(1) requires a DPP everywhere else to be subject to an "at least once" EIS: "At least once the Secretary shall declare the approval of a development and production plan in any area or region (as defined by the Secretary) of the outer Continental Shelf, other than the Gulf of Mexico, to be a major Federal action."[147] Likewise, the exception from the exception appears in subsection (l): "The Secretary may require the provisions of this section to apply to an oil and gas lease issued or maintained under this subchapter, which is located in that area of the Gulf of Mexico which is adjacent to the State of Florida."[148]

Whatever the legislative history may reflect about Congress's intentions with respect to additional NEPA review in the Western Gulf, the statutory language controls the proper interpretation of the OCSLA 1978 amendments. The statutory language was precise about what must happen outside the Western Gulf OCS region—mandatory DPPs for all leases and an EIS at least once rules the Department can now or could then have imposed in the Western Gulf were less obvious. But the statute did not prohibit NEPA review of Federal action there; it made the requirements of section 25 inapplicable there.

The statute also addresses EPs. The provisions codified at 43 U.S.C. § 1340(c) require lessees to submit EPs and require the Secretary to approve such plans within 30 days of submission. The statute does not address whether the plans qualify as major Federal actions significantly affecting the human environment, nor does it except or address the Gulf of Mexico.[149] In addition, the statute permits the Secretary to require lessees to submit an APD for a permit to drill in accordance with an approved EP.[150]

D. Departmental Manual Offshore Categorical Exclusions

At the time of the first publication of the Departmental Manual, responsibility for addressing Outer Continental Shelf activity fell to the USGS. The USGS proposed its NEPA compliance procedures, and also its first Departmental Manual CEs, in 1980.[151] From these first proposals, USGS identified all APDs as categorically excluded so long as they were described in an approved EP or DPP.

In the Western Gulf of Mexico, however, USGS also excluded the EP or DPP from the NEPA process.[152] The only explanation for this was a citation to USGS offshore rules implementing the OCSLA amendments of1978. Without articulating its logic, USGS thus suggested that the statute itself prohibited NEPA review of both DPPs and EPs in the Western Gulf of Mexico. At a minimum, USGS may have believed that construing section 25's exception for DPPs in that region in any other manner would provoke a controversy. USGS gave no reason for concluding that Western Gulf EPs would be categorically excluded. It proposed the following CEs:

> (24) Approval of an Application for Permit to Drill (APD) an OCS oil and gas exploration or development well when said well is described in an approved exploration plan, development plan, or production plan. . . .
>
> (36) Approval of OCS exploration plans in the Western Gulf of Mexico (30 C.F.R. § 250.2). . . .
>
> (37) Approval of OCS development/production plans in the Western Gulf of Mexico (30 C.F.R. § 250.2).[153]

Proposed CEs 27-30 addressed onshore APDs as CEs. Acknowledging that DPPs likely addressed action that would cause (according to Congress, at least) significant impacts, USGS noted that elsewhere such plans would likely require an EIS. USGS listed as "major actions normally requiring an EIS":

> Approval of an OCS oil and gas development and production plan in a frontier area, when the plan is declared to be a major Federal action in accordance with Section 25 of the OCS Lands Act Amendments of 1978 (30 C.F.R. § 250.34-4).[154]

The proposed USGS Departmental Manual provisions were approved as a final appendix to the Department's NEPA procedures on January 19, 1981, and published in the Federal Register on January 23, 1981.[155] Without further examining differences between the Western Gulf and other areas of the Outer Continental Shelf, USGS cited the OCSLA amendments of 1978 and 30 C.F.R. § 250. It did, however, reinforce the distinctions in response to comments. One commenter requested a change requiring EISs for all DPPs *except* in the Gulf of Mexico. USGS disagreed and stated that it would "follow the environmental review process established in 30 C.F.R. § 250.34."[156] USGS explained: "We will, in any area or region of the OCS, *except in the Western Gulf of Mexico*, prepare at least one development/production plan EIS pursuant to subsection 25(e) of the OCS Land[s] Act Amendments of 1978."[157]

The final appendix revised the order of CEs and separated onshore oil and gas CEs from offshore ones.[158] But in concept the offshore CEs were unchanged. The final CEs appeared in Part 2.4 of the USGS appendix to Chapter 6 of the Departmental Manual. The offshore CEs were as follows:

> (b) Approval of an OCS exploration or development/production plan in the Western Gulf of Mexico (30 C.F.R. § 250.2) which does not require an environmental report from an operator pursuant to item 3 of [Notice to Lessee] 80-6. . . .
>
> (d) Approval of an APD for an oil and gas exploration or development well when said well and appropriate mitigation measures are described in an approved exploration plan, development plan, or production plan.[159]

The Departmental Manual was amended in 1986, after the establishment of and transfer of oil and gas responsibilities to MMS. MMS addressed oil and gas development in Appendix 10 to Chapter 6, which was approved August 12, 1986. The Departmental Manual was reorganized again on May 27, 2004. Appendices produced by individual bureaus within the Department were assigned chapter numbers, and MMS's CEs were placed in Chapter 15. The critical elements of the original USGS CEs, however, never changed: An APD for any offshore well is categorically excluded from NEPA review so long as it is described in an approved plan for exploration, development, production, or operations; and, in the Western Gulf, such plans are themselves categorically excluded from the NEPA process. Other changes were significant, however. MMS separated the Western from the Central Gulf regions but defined plans submitted for offshore leases within both to be CEs. MMS added the newly created DOCD—the counterpart to the DPP required to be submitted by Gulf lessees—as a CE. And MMS revised the CEs for planning documents to ensure that exceptions were available in particular environmental situations. As currently written, the offshore CEs are as follows:

> (10) Approval of an offshore lease or unit exploration[,] development/production plan or a Development Operation Coordination Document in the central or western Gulf of Mexico (30 C.F.R. § 250.2) *except those proposing facilities: (1) In areas of high seismic risk or seismicity, relatively untested deep water, or remote areas; or (2) within the boundary of a proposed or established marine sanctuary, and/or within or near the boundary of a proposed or established wildlife refuge or areas of high biological sensitivity; or (3) in areas of hazardous natural bottom conditions; or (4) utilizing new or unusual technology*. . . .
>
> (12) Approval of an Application for Permit to Drill (APD) an offshore oil and gas exploration or development well, when said well and appropriate mitigation measures are described in an approved exploration plan, development plan, production plan, or Development Operations Coordination Document.[160]

The language italicized above is important. The addition of exceptions to the CEs for particular circumstances where environmental damage could be severe only makes sense if MMS necessarily recognizes that the basis for the CE is not a statutory mandate but rather the CEQ definition of a CE—an action not normally causing an effect on the quality of the human environment. If the reason for the CE is a mandatory statutory exclusion from NEPA, the statute cannot be overridden by a regulatory exception.[161] Previously, USGS had never suggested that approval of an EP or DPP met the definition of a CE. To the contrary, Congress defined approval of a DPP as something normally causing significant impacts that required an EIS at least once. This significant change in the language raises questions about the basis for the CEs.

It remains true that approval of plans outside portions of the Gulf of Mexico "normally" require an EIS. The OCSLA 1978 amendments continue to be the source of that Departmental Manual provision.[162] Nonetheless, the fact that MMS came to believe the CEs were founded in the CEQ and Departmental definitions was also reinforced elsewhere in the 2004 Departmental Manual, which specifies, consistent with the CEQ regulations and with 43 C.F.R. § 46, that a CE cannot apply to a proposed action if one of the exceptions to the application of CEs pertains.

The MMS CEs, reorganized in 2004, state that "the following MMS actions are designated categorical exclusions *unless the action qualifies as an exception under Appendix 2 of 516 [Departmental Manual]* 2."[163]

This statement is decidedly problematic. First, the statement would not be true, if (as USGS originally suggested) the CE was mandated by law.[164] Second, Chapters 1-6 of the Departmental Manual was revised in 2009 and condensed into four chapters. Nothing in the 2009 version of Chapters 1-4 defines a CE or specifies the exceptions to the application of CEs that were present in Chapter 2 in 2004 when MMS's Chapter 15 was created.

It is important to note that the remainder of the Departmental Manual, as amended, does properly refer back, as it must, to the superior application of the CEQ regulations and Departmental regulations at 43 C.F.R. § 46. Chapter 1 states that it "supplements the CEQ and DOI regulations and must be read in conjunction with both."[165] Elsewhere Chapter 1 specifies that in case of any discrepancies among Departmental guidance, "Departmental regulations at 43 CFR 46 and 516 Departmental Manual 1-4 shall govern."[166] MMS's Chapter 15, however, became more and more attenuated from its original purpose, MMS regulations, and the Departmental Manual itself. If, in fact, Department regulations govern in the case of a discrepancy between MMS regulations and the Departmental Manual, and the statute is *not* the source of the CEs for EPs, DPPs, and DOCDs, then it is questionable whether the CEs are valid in the absence of some reasonable determination showing that approval of such plans is "a category or kind of action that has no significant individual or cumulative effect on the quality of the human environment."[167]

E. Departmental Rules Implementing OCSLA Amendments of 1978

USGS promulgated rules applicable to offshore oil and gas operations that exempted lessees in the Western Gulf region from the requirement to prepare a DPP. In regulations incorporated into the CEs adopted by USGS and later MMS, codified at 30 C.F.R. § 250, the Department defined the Western Gulf of Mexico and addressed "compliance with [NEPA]."[168] In this latter regulation, which otherwise addressed NEPA review in association with EPs and DPPs, the Department made clear that USGS must "declare the approval of a development and production plan" in "the OCS, except in the western Gulf of Mexico, to be a major federal action at least once pursuant to subsection 25(e) of the Act."[169] This regulation was promulgated on September 14, 1979.[170] USGS's final 1981 CEs cross-referenced these rules.

Later, on several occasions, MMS redesignated and amended these regulations to incorporate extensive and specific requirements for the submission of EPs and DPPs.[171] These changes amended Part 250, which addresses whether the OCSLA amendments of 1978 prohibit the Department from requiring DPPs in the Western Gulf; accordingly. MMS created a "development operations coordination document," or DOCD, which was a substitute document required only in the Western Gulf region:

> Development and Production Plans are not required for leases in the western GOM.
>
> For these leases, the lessee shall submit to the Regional Supervisor for approval a Development Operations Coordination Document with all information necessary to assure conformance with the Act, other laws, applicable regulations, lease provisions, or as otherwise needed to carry out the functions and responsibilities of the Regional Supervisor.[172]

References to DPPs include these Gulf of Mexico DOCDs.[173] These regulations specified, as in the OCSLA 1978 amendments, that the Department would evaluate the environmental impacts of a DPP at least once in an EIS, except in the Western Gulf of Mexico. 174

In 2005, MMS again revised Part 250. Though MMS outlined several of its changes, it explained that the purpose of the amendments was to codify directives to lessees already in place in Notices to Lessees.[175] This rulemaking contained particular requirements for the contents of EPs, DPPS, and also DOCDs: consultation, environmental review, addressing of direct and cumulative effects on onshore and offshore environments resulting from implementation of EPs, DPPs, and DOCDs.[176]

In the preamble to this rulemaking, MMS addressed its NEPA processes, including the CEs, in a way indicating that MMS was unaware USGS had established the CEs by citing the OCSLA 1978 amendments; of the fact that DPPs must, at some point, be subject to an EIS; or of the fact that a CE is excepted from the NEPA review process rather than a part of it. MMS correctly explained that it "must comply with" NEPA, and relevant regulations.[177] But it went on to state, "According to NEPA requirements, MMS must prepare an [(EA)] in connection with its review of plans for activities on the OCS . . . . The appropriate MMS Region prepares these analyses for every plan received." However, the NEPA regulations (40 C.F.R. § 1508.4) do allow agencies to exclude categories of actions from the preparation of an EA or an EIS when agency procedures have demonstrated that these actions—individually or cumulatively—do not have a significant impact on the environment.

MMS follows the procedures outlined in the Departmental Manual[178] to categorically exclude routine OCS lease or unit plans in the Western and Central Gulf of Mexico areas unless certain exceptions are present. Some exceptions pertain to the nature of the proposed activity, and others to the nature of potential environmental impacts that may result from the activity. When MMS processes plans using a Categorical Exclusion Review, the agency reviews the proposed activity and the potential environmental impacts at the proposed site. These actions do not require MMS to prepare an EA, and MMS may limit the information that the lessee/operator is required to submit unless the information is required for compliance with other Federal laws. MMS prepares an EA in its review of plans that meet the criteria of the specified exceptions to the CE criteria or will prepare an EIS. Whether MMS reviews plans through the Categorical Exclusion Review or EA process, the agency requires that environmental impacts be avoided or diminished to an acceptable level through plan amendments or conditions MMS imposes during the plan approval process.[179]

Thus, by 2005, MMS was asserting that DOCDs submitted for portions of the Gulf are categorically excluded from NEPA review because they are "routine" and, by definition, not the sort of action that will have an impact on the quality of the human environment. At the same time, MMS believed that elsewhere planning documents normally require EAs (instead of the "at-least-once" EIS). In addition, MMS regulations at 43 C.F.R. § 250 addressed environmental requirements with respect to DOCDs, and treated them as the same as DPPs. For all planning documents, the regulations compelled submission of documents for purposes of NEPA review, reflecting the view that environmental impacts could result from the approvals. These positions about the nature of effects resulting from approval of planning documents cannot be squared.[180]

III. NEPA Issues Surrounding the Macondo Well

A. Question 1—Categorical Exclusions

MMS concluded that actions in approving the Exploration Plan and APD for the Macondo well fell within CEs, and therefore were not subject to NEPA. By CEQ's definition, these actions were therefore in a "category of actions which do not have a significant effect on the human environment and which have been found to have no such effect in procedures adopted by a Federal agency in implementation of the regulations (§ 1507.3) and for which, therefore, neither an [EA] nor an [EIS] is required. . . . ."[181] Because the idea that drilling a deepwater well on the Outer Continental Shelf is the sort of action that "does not have a significant effect on the human environment" appears, with hindsight, to be inherently illogical, the question arises whether BLM properly applied the CE procedures.

DOI's precise stance on these topics is unclear. Soon after enactment of the OCSLA 1978 amendments and the 1978 CEQ regulations, USGS created CEs in the Western Gulf. It categorically excluded from NEPA review all exploration, development, and production plans, and applications for permits to drill (so long as they were addressed in an approved plan). In doing so, USGS never identified such plans as actions, as defined by the CEQ or DOI, that are within the category of actions not having a significant impact on the human environment. Instead, by citing the OCSLA 1978 amendments, USGS implied that the law compelled CEs. That this happened in 1980-81 (as the Departmental Manual provisions moved from draft to final) is not surprising given the legislative history of the OCSLA 1978 amendments. Section 25 of the OCSLA1978 amendments, however, addressed DPPs but excepted the Western Gulf region from its application. This is not the same thing as saying that Federal action in that region would be exempt by law from NEPA review.

By regulation, MMS later created a new planning document for approval in the Gulf—a DOCD—thus seemingly acknowledging that section 25 of the OCSLA 1978 amendments did not define the Gulf as excluded from the regulation requiring planning documents. If that is the case, by analogy it seems unlikely that NEPA exemption for the region is compelled by section 25.

As the years passed, MMS appeared to lose sight of how the CEs came into being. Without questioning whether they *were* properly defined as CEs, MMS presumed that they were properly defined, and applied exceptions to the CEs in locations where particular types of environmental effects might be felt. Thus, MMS came to act as if the CEQ and Departmental definitions of an action, which do not normally have an effect on the human environment, properly applied to approvals of EPs, DPPs, and DOCDs in the relevant Gulf regions, while enforcing regulations that made clear that the same approvals in other regions could not meet that definition and required EAs and EISs.[182] It seems difficult to sustain such competing views of the environmental impacts of the same planning documents based solely on lines drawn on a map. MMS has never precisely addressed the justification for its employment of CEs. By employing terms such as "Categorical Exclusions Review" for the process to decide whether an agency action should fall within a CE found in the Departmental Manual, the inference could be drawn (and may have been intended by MMS) that the Categorical Exclusion Review was a third level of NEPA review rather than a plain *exemption from* NEPA review—the quintessential definition of a CE in CEQ's regulations.[183]

With respect to the Macondo well, MMS conducted NEPA review at the lease planning stage and the lease sale stage. Relevant to the use of CEs, however, MMS considered an EP for Lease 206 (the lease on which the Macondo well was drilled), an amended EP, and four APDs. In every case, MMS applied CEs.[184] There are several problems with these decisions by MMS.

First, it is not clear why, 29 years into the consistent agency declaration of CEs, these leases were defined as CEs in the first place. The rationale behind the CE for the exploration plans was not explained. While it has been suggested that USGS categorically excluded EPs in the Gulf because of the 30-day approval requirement of the OCSLA,[185] this time frame applies throughout the Outer Continent Shelf under that statute. Yet, EPs in other regions are subject to NEPA under 43 C.F.R. § 250, as discussed by MMS in the preamble to its 2005 rules. The finding that the Gulf region EPs are exempt from NEPA review seems to be inexplicable.

Second, to the extent that the agency came to believe the OCSLA 1978 amendments did not compel the CEs, there seems to be no basis for the implicit conclusion that approvals of EPS, DPPs, or DOCDs in the Gulf region are actions that do not have an impact on the human environment. The same planning documents were found to require NEPA review in agency rules at 30 C.F.R. § 250, and, therefore, without a sensible explanation for the difference, the continued employment of CEs in the Gulf region on this basis is suspect.

Third, the APDs are subject to CEs when the well to which they pertain is subject to an approved planning document. While it does not constitute NEPA review, a CE is a product of NEPA and CEQ regulations. The fact that an action is the subject of another NEPA document, and therefore does not require duplicate NEPA review, is an effect of tiering. Defining APDs as categorically excluded from NEPA (except in the context of Energy Policy Act CEs, which are identified by law) is a questionable extension of the tiering concept. Further, to the extent that defining approval of APDs as categorically excluded derives from the notion that it is unnecessary to do another NEPA review because one was already done for the planning document, this logic does not and cannot apply when the planning document was not subject to NEPA in the first place. No NEPA review can be tiered to a non-NEPA document, and a CE is not a NEPA document. Accordingly, application of the CEs for APDs to any geographical place, though long-standing, has inherent logical flaws; application in the Western Gulf of Mexico where the planning documents were never subject to NEPA review at all would seem unsustainable.

Fourth, even assuming that the planning documents were properly defined as categorically excluded from NEPA review, and the CEQ and Departmental regulations were the source for their designation as CEs, it follows that MMS was obligated by regulation to question whether any of the exceptions applied based on extraordinary circumstances.[186] Without reproducing the list of exceptions, it does not seem clear that MMS seriously undertook the task of examining this list carefully. The notion that the largest oil spill in American history resulted from a blowout on a well that had been defined by the agency as exempt from NEPA because it is an action "which do[es] not have a significant effect on the human environment"[187] indicates that the application of the CEQ definition was off-kilter and the exceptions were not sufficiently examined. The concern goes beyond after-the-fact logic. The exceptions to the application of a CE include actions which "[h]ave highly uncertain and potentially significant environmental effects or involve unique or unknown environmental risks."[188] The May 2000 Gulf of Mexico Deepwater Operations and Activities EA expressly discussed the potential for a deepwater spill and the length of time needed to control one (60-120 days); it concluded that "further investigation is needed" to evaluation the consequences of a deepwater blowout.[189] Given the actual events after the blowout took place, it is clear that such investigation was never completed. How a deepwater EP escaped NEPA review under the exception to CEs is difficult to understand.

B. Question 2—Consultation

Since the blowout, much has been made of the fact that MMS did not fully appreciate or address comments of sister or local agencies in the NEPA process. Although MMS was not obligated to agree with other agencies' comments, MMS was obligated to respond in the EISs to which the Macondo well approvals ultimately were tiered, and to "assess and consider comments . . . and respond" in one of the manners required in 40 C.F.R. § 1503.4. The question of whether MMS properly considered agency and public comments with respect to the lease sale EA should be answered on a case-by-case basis to determine whether MMS considered comments to the extent practicable.[190] In both cases, to the extent MMS addressed comments, its response was required to be well-reasoned, and could not be arbitrary.

The basis for much of the concern regarding the adequacy of consultation and coordination stems, as with the concern for the adequacy of NEPA review at all, from the employment of CEs for the EPs and APDs for the Macondo well. MMS's general regulations regarding planning documents are nonspecific as to the kind of NEPA review anticipated and do not clarify how consultation will occur, instead leaving that question of compliance with the CEQ and Departmental NEPA rules. In addition, the regulations call for documents in association with DPPs and DOCDs that are not, by rule, provided to the public, and so, even with NEPA compliance, it is unclear what the public or local agencies may see in the context of NEPA review. It does seem, however, that, to the extent the agency considered the action to be categorically excluded from NEPA review, no NEPA compliance would take place at all that would effectuate consultation. Accordingly, the question of MMS's compliance with consultation requirements in the context of CEs is strictly derivative. If the use of the CE was appropriate, then consultation in association with NEPA review is not at issue because there was no NEPA review.

MMS regulations governing the planning documents presume the applicability of all relevant laws, including NEPA, and do not address the CEs for planning documents in the Gulf. They do not precisely duplicate NEPA regulations, either those prepared by CEQ or DOI. For example, with respect to environmental compliance, the regulations require the lessee to submit an environment impact analysis (EIA),[191] for purposes of allowing MMS to "comply[] with the National Environmental Policy Act (NEPA) of 1969 and other relevant Federal laws."[192] With respect to consultation, the regulations address the *lessee's* obligation to inform MMS of the lessee's consulting history: "Your EIA must include a list of agencies and persons with whom you consulted, or with whom you will be consulting, regarding potential impacts associated with your proposed exploration activities."[193] The requirements for DPPs and DOCDs are the same. The lessee must submit an EIA for these planning documents,[194] for purposes of allowing MMS to comply with, *inter alia*, NEPA.[195] And, as with EPs, the regulations specify that the lessee must describe its own consultation activities with respect to DPPs or DOCDs.[196] In addition, deepwater operators must submit a "deepwater operations plan" that "provides sufficient information for MMS to review a deepwater development project, and any other project that uses non-conventional production or completion technology."[197] The deepwater operations plan must be accompanied by a "Conceptual Plan" for the project.[198] The regulations at 30 C.F.R. § 250.286 through 250.295 detail the requirements for these documents, but contain no requirements pertaining to notification of the public, consultation with local agencies, or compliance with NEPA. The deepwater operations plan and Conceptual Plan do not substitute for but do supplement the DPP or DOCD. Similarly, the lessee submitting a DPP or DOCD must also submit a conservation information document.[199] Once again, these regulations do not describe NEPA or consultation.

From these regulations, one must surmise that MMS, when preparing appropriate NEPA reviews for EPs and DPPs or DOCDs, will conduct appropriate consultation. Much information, however, is submitted in association with DPPs and DOCDs—with the deepwater operations plans and conservation information documents—that is not, by rule, publicized to interested parties, and thus would only be publicly available if provided in association with a properly-prepared NEPA review document. If the DPP or DOCD is only considered as an EA, such information would be provided "to the extent practicable." By contrast, material that is a part of the DPP or DOCD is subject to specific review and consultation provisions of the MMS operations rules.

This discussion of rules relating to planning documents, and associated NEPA review, is only of tangential interest in the Western Gulf, where the agency employs CEs for their approval. The MMS operations regulations are silent about CEs, or how their use affects MMS's decisions with respect to consultation. In the preamble to the 2005 rulemaking, MMS explained that it might choose to limit the material the lessee would otherwise be required to submit if MMS decided that a plan approval is categorically excluded. In the case of a CE, "MMS may limit the information that the lessee/operator is required to submit unless the information is required for compliance with other Federal laws."[200] Accordingly, in the case of a CE, it is not entirely clear that MMS would even demand compliance with requirements of the operations rules related only to NEPA.

In conclusion, because consultation with respect to an EA or EIS is governed by NEPA rules promulgated by CEQ and the DOI, and is only necessary "to the extent practicable" for EAs, it is not at all clear what information is normally available to the public or to appropriate local agencies, how it is disseminated, or whether there is an opportunity to consult or comment in association with an EA prepared for a planning document. But in the Western Gulf of Mexico, where planning documents are categorically excluded by Departmental Manual, as indicated in Notices to Lessees, no NEPA consultation would take place at all, and MMS may, by Notice to Lessees, limit the information required to be submitted by the lessee.

C. Question 3—Reasonably Foreseeable Impacts

The NEPA documents that sought to examine the risk of an oil spill considered the possibility of a blowout the size of the 2010 event to be so unforeseeable that these documents did not address it. Although MMS did not examine a type of blowout of the nature that occurred as a "worst case scenario" in the relevant NEPA documents, BP apparently did look at a worst case discharge for the purpose of its Oil Spill Response Plan, which is an appendix to the EP.[201] In this document, BP addressed a potential spill of up to 300,000 barrels per day. Possibly because the EP to which the Oil Spill Response Plan was appended was identified as a CE, this projection was never included in any NEPA study by MMS. As a purely NEPA question, then, this paper asks whether sufficient information was available to justify a more detailed analysis by MMS of a large-scale oil spill in relevant NEPA analyses on grounds that such an impact was reasonably foreseeable.

Several factors undercut MMS's decision to address potential spills in its NEPA documents (Programmatic EIS, Multi-Sale EIS, and Lease Sale 206 EA), all the while eschewing any discussion of a blowout of a significant nature before the blowout took place. To the extent such a discussion would be found to be legally required, the location of the discussion would have been in an EIS to which subsequent documents are tiered. Conversely, a FONSI for an EA, such as the Lease Sale 206 EA, may have been unwarranted if a spill of the size projected in the Oil Spill Response Plan had been addressed in an EA. If a blowout of this magnitude was reasonably foreseeable for purposes of the Oil Spill Response Plan, it is hard to imagine how such a spill would not be a reasonably foreseeable impact to be examined under NEPA.[202] MMS appears to have relied on geographic boundaries (in this case a geopolitical boundary) for classifying the foreseeability of impacts of the Ixtoc spill off the coast of Mexico in 1979. Whether a repeat event was reasonably foreseeable would depend on an explanation why what happened in 1979 could not happen again, or why it was too "speculative" or unduly remote.[203]

This question is now academic. Whatever was true about what was reasonably foreseeable to consider by way of effects before the Macondo blowout, it would be arbitrary hereafter for MMS to undertake NEPA review without considering the sort of effects that resulted from the Macondo well blowout as reasonably foreseeable. The question of what is reasonably foreseeable has two components. What must the agency consider? And does it compel an EIS?[204] While it can be debated whether the likelihood of another blowout of the magnitude of the Deepwater Horizon spill is so small that an EIS is unnecessary, this might be a hard sell absent the inclusion of specific mitigation measures capable of reducing the risk below that necessary to trigger an EIS. Absent such mitigation, MMS will likely be required at some stage of its decision-making process to prepare an EIS of some nature that considers such effects, to which other NEPA documents may be tiered.

End Notes

[1] Vermont Yankee Nuclear Power Corp. v. Natural Resources Defense Council, Inc., 435 U.S. 519, 558 (1978).

[2] Kleppe v. Sierra Club, 427 U.S. 390, 410 n.21 (1976).

[3] Baltimore Gas & Electric Co. v. Natural Resources Defense Council, Inc., 462 U.S. 87, 97 (1983).

[4] Robertson v. Methow Valley Citizens Council, 490 U.S. 332, 350-51 (1989).

[5] 42 U.S.C. § 4321.

[6] *See* Robertson, 490 U.S. at 349.

[7] *Id.* at 350-51 (citations, footnote omitted); Marsh v. Oregon Natural Resources Council, 490 U.S. 360, 371 (1989).

[8] 42 U.S.C. § 4332(2).

[9] 42 U.S.C. § 4332(2)(C).

[10] 42 U.S.C. § 4332(2)(C)(i)-(v).

[11] 42 U.S.C. § 4332(2)(E).

[12] *See, e.g.*, Sabine River Authority v. U.S. Department of the Interior, 951 F.2d 669, 675 (5th Cir. 1992), *citing* Metropolitan Edison Co. v. People Against Nuclear Energy, 460 U.S. 766 (1983) (NEPA is triggered by "physical change in the environment").

[13] 40 C.F.R. § 1500 *et seq.*

[14] 40 C.F.R. § 1500.3.

[15] 40 C.F.R. § 1500.2(b); 1507.3(a).

[16] 40 C.F.R. § 1501.3; 1501.4.

[17] 40 C.F.R. § 1508.11.

[18] 40 C.F.R. § 1508.9.

[19] *See* 40 C.F.R. § 1507.3(b).

[20] 40 C.F.R. § 1508.13.

[21] *Id.*

[22] *Id.*

[23] *See* 40 C.F.R. § 1508.20.

[24] 40 C.F.R. § 1508.9(a)(3).

[25] 40 C.F.R. § 1508.10.

[26] 40 C.F.R. § 1508.22.

[27] 40 C.F.R. § 1501.4(a)(2).

[28] 40 C.F.R. § 1502.

[29] *Id.* C.F.R. § 1502.19 and 1502.25.

[30] 40 C.F.R. § 1508.9(b).

[31] 42 U.S.C. § 4332(2)(C) (EIS required when significant impacts are found).

[32] 40 C.F.R. § 1502.16 (EIS includes discussion of "adverse environmental effects which cannot be avoided").

[33] 490 U.S. at 352-53 (footnote and citation omitted).

[34] 5 U.S.C. §§ 704-06.

[35] Marsh, 490 U.S. 360; *see also* 40 C.F.R. § 1502.24 (agencies must "insure the professional integrity, including scientific integrity" of analyses in EISs).

[36] 42 U.S.C. § 4332(2)(C), (E); 40 C.F.R. § 1502.1; Kleppe, 427 U.S. at 410 n.21; Natural Resources Defense Council, Inc. v. Hodel, 819 F.2d 927, 929 (9th Cir. 1987).

[37] State of California v. Block, 690 F.2d 753, 761 (9th Cir. 1982), *quoting* Trout Unlimited v. Morton, 509 F.2d 1276, 1283 (9th Cir. 1974); Scientists' Institute for Public Information v. Atomic Energy Commission, 481 F.2d 1079, 1092 (D.C. Cir. 1973).

[38] 5 U.S.C. § 706.

[39] Cabinet Mountains Wilderness v. Peterson, 685 F.2d 678, 681-82 (D.C. Cir. 1982); Maryland-National Capitol Park & Planning Commission v. U.S. Postal Service, 487 F.2d 1029, 1040 (D.C. Cir. 1973).

[40] 40 C.F.R. § 1508.10.

[41] 40 C.F.R. § 1500.4 ("Reducing paperwork").

[42] 40 C.F.R. § 1500.4(p) (*citing*40 C.F.R. § 1508.4).

[43] 40 C.F.R. §1507.3.

[44] 40 C.F.R. § 1508.4.

[45] 951 F.2d at 679.

[46] 40 C.F.R. § 1500.4(p).

[47] 40 C.F.R. § 1507.3(b)(2)(ii).

[48] 40 C.F.R. § 1502.16; 1502.14.

[49] 40 C.F.R. § 1508.7.

[50] 40 C.F.R. § 1502.16(h).

[51] 40 C.F.R. § 1508.9(b).

[52] 40 C.F.R. § 1508.8; 1508.25(c).

[53] 40 C.F.R. § 1508.8(a).

[54] 40 C.F.R. § 1508.8(b).

[55] Metropolitan Edison, 460 U.S. at 774-75.

[56] 40 C.F.R. § 1508.7.

[57] *Id.*

[58] 40 C.F.R. § 1502.22.

[59] *Id.*

[60] 51 Fed. Reg. 15624.

[61] *Id.*

[62] 490 U.S. at 356.

[63] 40 C.F.R. § 1502.22(b)(1).

[64] *Id.*

[65] 40 C.F.R. § 1500.2(d).

[66] 40 C.F.R. § 1501.6.
[67] 40 C.F.R. § 1506.6(a).
[68] 40 C.F.R. § 1501.6.
[69] 40 C.F.R. § 1501.6(b).
[70] 40 C.F.R. § 1501.7.
[71] *Id.* (*see* 1507.3(e) (notice may be issued closer to the time the EIS is prepared)).
[72] 40 C.F.R. § 1501.7(a).
[73] 40 C.F.R. § 1502.9(a).
[74] *Id.*
[75] 40 C.F.R. § 1503.1.
[76] 40 C.F.R. § 1503.1(a)(1).
[77] 40 C.F.R. § 1503.1(a)(2)-(4).
[78] 40 C.F.R. § 1503.1(a)(4).
[79] 40 C.F.R. § 1503.4(a).
[80] *E.g.*, Center for Biological Diversity v. U.S. Forest Service, 349 F.3d 1157, 1168-69 (9th Cir. 2003).
[81] Life of the Land v. Brinegar, 485 F.2d 460, 472 (9th Cir. 1973).
[82] Center for Biological Diversity, 349 F.3d at 1169.
[83] California v. Block, 690 F.2d 753, 771 (9th Cir. 1982).
[84] 40 C.F.R. § 1501.4(b).
[85] 40 C.F.R. § 1501.4(e)(1).
[86] 40 C.F.R. § 1501.4(e)(2).
[87] *Id.*
[88] 341 F.3d 961, 970 (9th Cir. 2003).
[89] 314 F.3d 1006, 1016 (9th Cir. 2002).
[90] Alliance to Protect Nantucket Sound, Inc. v. U.S. Department of Army, 398 F.3d 105, 115 (1st Cir. 2005).
[91] Greater Yellowstone Coalition v. Flowers, 359 F.3d 1257, 1279 (10th Cir. 2004).
[92] 306 F.3d 1235, 1238-39 (2d Cir. 2002)
[93] 281 F. Supp. 2d 209, 226-27 (D.D.C. 2003)
[94] Biodiversity Conservation Alliance v. U.S. Bureau of Land Management, 404 F. Supp. 2d 212, 220 (D.D.C. 2005).
[95] 404 F. Supp. at 220, *citing* 40 C.F.R. § 1501.4(b).
[96] Environmental Defense Fund v. Andrus, 596 F.2d 848, 852-53 (9th Cir. 1979).
[97] Mobil Oil Corp. v. FTC, 562 F.2d 170, 173 (2nd Cir. 1977); Sierra Club v. Peterson, 717 F.2d 1409, 1414-15 (D.C. Cir. 1983).
[98] 40 C.F.R. § 1502.9(c)(1).
[99] *Cf.* 42 U.S.C. § 4332(2)(C).; Marsh v. Oregon Natural Resources Council, 490 U.S. at 373-74 (footnotes omitted).
[100] 40 C.F.R. § 1508.28.
[101] 284 F.3d 1062, 1074 (9th Cir. 2002).
[102] Kern, 284 F.2d at 1074.
[103] 377 F.3d 1147, 1162 (10th Cir. 2004).
[104] *Id.*
[105] The U.S. Forest Service uses a "validation and verification" procedure.
[106] On June 18, 2010, Secretary of the Interior Ken Salazar ordered that the Minerals Management Service be officially renamed the Bureau of Ocean Energy Management, Regulation, and Enforcement (BOEMRE). For consistency, throughout this paper, we will

refer to the agency as the Minerals Management Service. As of the writing of this paper, the Departmental Manual has not been changed to acknowledge the BOEMRE.

[107] 43 C.F.R. § 46.120(d) ("Responsible Officials should make the best use of existing NEPA documents by supplementing, tiering to, incorporating by reference, or adopting previous NEPA environmental analyses to avoid redundancy and unnecessary paperwork.")

[108] 43 C.F.R. § 46.135; 46.140.

[109] 40 C.F.R. § 46.140(c) (first italics added; second in original).

[110] 43 C.F.R. § 46.155.

[111] *Id.*

[112] 43 C.F.R. § 46.435.

[113] *Id.*

[114] 43 C.F.R. § 46.305(c).

[115] 43 C.F.R. § 46.310; 46.315; 46.320; 46.325.

[116] 43 C.F.R. § 46.400; 46.405; 46.415; 46,420; 46.425.

[117] 43 C.F.R. § 46.205.

[118] 43 C.F.R. § 46.205(a).

[119] 43 C.F.R. § 46.205(c)(1).

[120] 43 C.F.R. § 46.215.

[121] 43 C.F.R. §. 46.205(d).

[122] 11 Departmental Manual 1.1.

[123] 11 Departmental Manual 1.2.

[124] 44 Fed. Reg. 40436.

[125] 45 Fed. Reg. 27541.

[126] 45 Fed. Reg. 27542.

[127] 516 Departmental Manual 2.3.A(1) (1980). 44 Fed. Reg. 40439 (July 10, 1979) (copied in final at 45 Fed. Reg. 27544 (Apr. 23, 1980)).

[128] 516 Departmental Manual 2.3.A(1) (1980).

[129] *Id.* at 2.3A(2).

[130] *Id.* at 6.4.A(3).

[131] Ultimately, MMS created three Gulf of Mexico areas: Western, Central, and Eastern. For purposes of NEPA, MMS grouped the Western and Central areas together.

[132] This paper does not separately document the extensive Notices to Lessees that have applied to CEs.

[133] *See* Public Law 95-372, section 25 at subsections(a)(1), (b), and (e).

[134] *Id.* at section 25(l).

[135] Report No. 95-284, the Subcommittee on Energy and Natural Resources Report to Accompany S. 9 (June 21, 1977) at 50.

[136] *Id.*

[137] *Id.* at 82.

[138] *Id.* at 83.

[139] *Id.* at 82.

[140] *Id.* at 62.

[141] *Id.*

[142] *Id.* at 144-45.

[143] *Id.* at 83.

[144] *Id.* at 154-55.

[145] *Id.* at 115-16.

[146] 43 U.S.C. § 1351(b).

[147] 43 U.S.C. § 1351(e)(1).
[148] 43 U.S.C. § 1351(l).
[149] *Id.*
[150] 43 U.S.C. § 1340(d).
[151] 45 Fed. Reg. 75336 (Nov. 14, 1980).
[152] *See* 516 Departmental Manual 2.4.A(2) (1980).
[153] 45 Fed. Reg. 75336-37 (CEs listed in the Chapter 6, draft USGS appendix Part 2.4).
[154] Draft USGS appendix Part 2.3, 45 Fed. Reg. 75336 (Nov. 14, 1980).
[155] 46 Fed. Reg. 7485.
[156] *Id.*
[157] *Id.* (emphasis added).
[158] Part 2.4.B(2)(a)-(h), at 46 Fed. Reg. 7487 (Jan. 23, 1981) (onshore CEs). Congress subsequently created a "rebuttable presumption" that certain *onshore* APDs are categorically excluded from NEPA review in section 390 of the Energy Policy Act of 2005, Pub. L. No. 109-58, 119 Stat. 594. 42 U.S.C. § 15942(a) and (b). The presumption exists generally if the wells are foreseeable in a NEPA document prepared in the previous five years.
[159] Appendix 5 at 2.4.B(3) (1981), published at 46 Fed. Reg. 7487 (Jan. 23, 1981).
[160] 516 Departmental Manual 15.4.C (2004 and 2010).
[161] *See* 43 C.F.R. § 46.205(d).
[162] 516 Departmental Manual 15.4 (2004 and 2010) ("Major Actions Normally Requiring an EIS").
[163] Emphasis added.
[164] 43 C.F.R. § 46.205(d).
[165] 516 Departmental Manual 1.1 (2010) (Purpose).
[166] 516 Departmental Manual 1.8.C (2010) (Mandate).
[167] 43 C.F.R. § 46.205.
[168] 30 C.F.R. § 250.34-4.
[169] 30 C.F.R. § 250.34-4(d).
[170] 44 Fed. Reg. 53700.
[171] *See* 63 Fed. Reg. 29479, 29485 (May 29, 1998).
[172] 30 C.F.R. § 250.204(d)(1) (2000).
[173] 43 C.F.R. § 250.200.
[174] 43 C.F.R. § 250.204(j).
[175] 70 Fed. Reg. 51478 (Aug. 30, 2005).
[176] *E.g.*, 30 C.F.R. § 250.227 (EPs).
[177] 70 Fed. Reg. 51478 (Aug. 30, 2005).
[178] 516 Departmental Manual 15 (2010).
[179] *Id.* (emphasis added).
[180] The Congressional Research Service's June 4, 2010 Report on the Deepwater Horizon Oil Spill quotes MMS as stating that CEs are available when "the impacts from the common operations are expected to be negligible to nonexistent based upon general information gathered during past environmental analysis." Congressional Research Service, *The 2010 Oil Spill: The Minerals Management Service (MMS) and the National Environmental Policy Act (NEPA)* at 12 (June 4, 2010). The source of the cited MMS document at CRS's Appendix C is unclear. To the extent the quote is from an MMS paper or statement, it reflects some confusion about the proper definition of a CE. If, in fact, MMS found during regular NEPA document review that a particular type of agency action routinely had

"negligible/non-existent" impacts, it would have been necessary for MMS to officially define such action as a CE based on the CEQ definition as a direct result of such experience with NEPA review, or else continue to perform NEPA review, which presumably would result in FONSIs. But no such event occurred. The CEs for DPPs and EPs were put in place for the Western Gulf region almost 30 years ago, when the first NEPA Departmental Manual was established.

[181] 40 C.F.R. § 1508.4.

[182] 30 C.F.R. § 250. *See also* MMS , *Oil and Gas Leasing, Exploration & Development Process*, Figure 1 (depicting EAs and EISs for EPs and DPPs).

[183] 40 C.F.R. § 1500.4(p).

[184] *See* Council on Environmental Quality, *Report Regarding the Minerals Management Service's National Environmental Policy Act Policies, Practices, and Procedures as They Relate to Outer Continental Shelf Oil and Gas Exploration and Development CEQ*, Figure 2 (Aug. 16, 2010) [hereinafter CEQ Report].

[185] 43 U.S.C. § 1340.

[186] 43 C.F.R. § 46.205; 46.210; 46.215.

[187] 40 C.F.R. § 1508.4

[188] 43 C.F.R. § 46.215(d). Though MMS employed CEs for the EP, MMS's operational regulations defined deepwater development as "non-conventional." 30 C.F.R. § 250.286.

[189] CEQ Report at Chapter II, pages 16-17.

[190] 40 C.F.R. § 1501.4(b).

[191] 30 C.F.R. § 250.227.

[192] 30 C.F.R. § 250.227(a)(3).

[193] 30 C.F.R. § 250.261(d).

[194] 30 C.F.R. § 250.261.

[195] 30 C.F.R. § 250.261(a)(3).

[196] 30 C.F.R. § 250.261(d).

[197] 30 C.F.R. § 250.286(a).

[198] 30 C.F.R. § 250.286(b).

[199] 30 C.F.R. § 250.296-250.299.

[200] 70 Fed. Reg. 51478 (Aug. 30, 2005).

[201] *See* 30 C.F.R. § 250.213. This rule requires a lessee to append the following to its EP: "Blowout scenario. A scenario for the potential blowout of the proposed well in your EP that you expect will have the highest volume of liquid hydrocarbons. Include the estimated flow rate, total volume, and maximum duration of the potential blowout. Also, discuss the potential for the well to bridge over, the likelihood for surface intervention to stop the blowout, the availability of a rig to drill a relief well, and rig package constraints. Estimate the time it would take to drill a relief well."

[202] 40 C.F.R. § 1508.8(b).

[203] 40 C.F.R. § 1502.22; Metropolitan Edison, 460 U.S. at 774-75.

[204] Inland Empire Public Lands Council v. Schultz, 992 F.2d 977, 981 (9th Cir. 1988) ("Both connected actions and unrelated but reasonably foreseeable future actions may result in cumulative impacts that can trigger an EIS."); Save the Yaak Comm. v. Block, 840 F.2d 714, 721 (9th Cir. 1988).

INDEX

A

B

C

F

G

H

I

J

L

M

N

O

P

R

S